GEOMORPHOLOGY OF DAMODAR BASIN

GEOMORPHOLOGY OF DAMODAR BASIN

S. N. P. Gupta

RAJESH PUBLICATIONS
New Delhi - 110 002

ISBN : 81-85891-39-7

Published by :
M. L. Gupta
Rajesh Publications
1, Ansari Road, Darya Ganj,
New Delhi - 110 002

Typesetting by :
Sanjay Enterprises
New Delhi- 110 002

Printed in India at :
A.P. Offset Printing Press, Delhi.

Ist Published in India : 2004

RAJESH PUBLICATIONS
Rs. 500.00

Preface

The author had two main goals in his mind while writing the present volume, Geomorphology of the Damodar Basin in Jharkhand. The first was to explain morphological variations in the area under study critcally and genetically; and the second was to evaluate the application of such a study to human aspects.

To achieve the goal, therefore, the work is divided into thirteen chapters. The first chapter introduces the subject conceptually. It also introduces the area. Previous studies, sources of information and methods of approach have also been included in this chapter.

The study of morphological variations and the landscape development requires knowledge of geology and structural characteristics of the area. Chapters second and third, therefore, deal with the geological formations, structural characteristics and tectonic features of the region. The fourth chapter describes the physiography of the Basin in general.

Geomorphological study has become quantitative. In the fifth, sixth and seventh chapters, therefore, morphometric techniques have been applied for the quantitative expressions of the linear, areal and relief aspects of the Basin.

Geomorphological interpretations of some of the significant features have been included in Chapter 8. Surface history and evolution of the drainage system have been done in Chapter 9.

In the tenth chapter attempts have been made to discover how the entire region came to appear in its present form. The geomorphic model which is applicable to this area to explain its varying morphological features has been examined. The eleventh chapter discusses the extent and role of the climatogenetic geomorphology in the geomorphic interpretation of the landforms of the area. An attempt has also been made *pari passu* to examine the influence of tectogenetic geomorphology in the evolution of the landforms of the Basin.

On the basis of the observations of the geomorphic features and other details the Basin has been divided into geomorphic regions. The discussion of each geomorphic region is included in the twelth chapter.

Chapter 13 discusses the application of geomorphology to human aspects of the area. The inclusion of this chapter is due to the fact that geomorphic aspects have great role on cultural landscape, e.g. the study of soil and soil erosion, hydrology, agricultural landscape, communication and settlement, etc. The analysis of these aspects requires the knowledge of geomorphic features of the area and thus, this chapter is an integral part of the present volume.

Facts have been illustrated through maps, tables, and photographs. An exhaustive bibliography is appended at the end of the thesis.

Thus, all efforts have been made to achieve the above mentioned goals as well as to make a humble contribution to the field of geomorphology which be judged only by the readers.

Shiva Narayan Prasad Gupta

Acknowledgement

The author is highly grateful to his teacher Dr. Enayat Ahmad, Professor and Head (Retd.) Department of Geography, Ranchi University, Ranchi, for his supervision.

The author also takes this oppurtunity of expressing his thanks to the University Grants Commission, New Delhi, for the award of the Teacher Fellowship.

The author would like to place on record his deep sense of gratitude to Dr. N. Prasad of Burdwan University, Prof. S. S Sinha, Giridih College and Dr. A. K. Sarkar, D. V. C. Hazaribag who accompanied the author in field work.

The author wishes to express his thanks to Dr. Kamla Prasad, Dr. A. Mahto and Dr. S. Sahu whose constructive criticism and help have always been valuable.

The author wishes to express his deep sense of gratitude to Smt. Veena Gupta who, inspite of many family problems, encouraged the author to undertake and complete the work.

The author would like to extend his deep sense of gratitude to his friends and officials (especially of D.V.C.).

Finally hearty thanks are also due to Sri. M. L. Gupta, proprietor, Rajesh Publications, New Delhi for the help and encouragement the author recieved to get the work published.

Contents

List of Maps and Diagrams

List of Plates

1

Introduction

1.1. Conceptual Aspects :

Geomorphology is the science which studies the landsurface of any part of the earth in relation to geogical structure, tectonic history and erosional forces. Thus, geomorphology is essentially an explanatory subject emphasizing the interpretation of the upper surface of land areas in relation to the different causes which have shaped it.

Some geomorphologists such as Sparks, hold that geomorphology is confined to the study of only the landforms of destruction or degradation and the depositional landform "as a whole is not within the scope of Geomorphology " (Sparks, 1972:1). This, however, appears to be unreasonable because many landforms are due to deposition.

There may be some overlapping between the meaning of such terms as physical feature, relief, topography, orography, bathy-orography, physiognomy, physiography and geomorphology. Physical feature generally refers to the topographic feature of upland or oceanic areas and there is always the reference to the mean sea level with which the surface features of land areas or the marine areas are compared. Relief is also generally used in the same sense as physical features but we know that its precise meaning is the difference between the highest and the lowest points of the particular area. The term orography or bathyorography is used in the same sense as physical features or relief and methods of representation are more or less the same. But where orography generally describes "the relief mountain systems" (Monkhouse, 1972:250), bathy - orography depicts "both the altitude of the land and the depths of the sea" (Monkhouse, 1972:36). Physiognomy is a term used generally by Human Geographers and is more or less identical to topography.

In both these concepts, in physiognomy and topography there is stress on the actual representation of landforms through cartographic techniques and it is attempted to obtain something more than by mere contour or chromatic or similar other methods. But probably in all these terms mentioned above, the chief object is to depict the surface and to describe it and no attempt is made to explain the causes of the landforms. But, when we study physiography the descriptive stage ends and explanatory stage starts. Some geographers and geologists have studied physiography in such detail and in such reasoned fashion that there is little difference between physiography and geomorphology.

Physiography, probably, represents an earlier stage in the history of geomorphology, where attempts were normally confined to the relation between structure and surface features or physiographic processes and landforms.

Under the name 'Geomorphology' the subject has assumed a wider meaning. In addition to geology, structure and processes emphasis is now given on the tectonics, eustatic changes of sea level, isostatic adjustments and other physical phenomena. Moreover physiography was merely a part of physical geography, geomorphology has developed almost as an independent subject because of its great application to human activities. Although, geomorphology is important in modern age, it was likely to be somewhat neglected. The reason is that it belongs neither wholly to geology nor entirely to geography nor exclusively to any other branches of science. It is a border - line subject almost shared half and half between geography and geology. Geology furnishes the information on physical and chemical composition of rocks, their ages and structural features. It also provides a clue to the succession in the geological formation. On the other hand geography observes the surface and its features at present, relates the characteristics of the surface to the characteristic of rocks as mentioned above. It also observes and examines the relation between the surface features and the processes of weathering, erosion and deposition. Thus, geomorphology cannot stand without geography and geology, sustaining it together.

Now, we can come to the problem of the utility of the subject. The purpose of geomorphology is not only to describe the morphology of the surface of the earth but also to explain its variations. The approach is genetic. This itself is very difficult as well as important discipline. Unless we know the causes of the landforms we cannot forsee what changes these landforms will undergo. Unless we know how these have behaved in recent past we cannot forecast how they will behave in future. As a specific example we are concerned with gully erosion which has devoured about 1.2% of the total surface of the country (Geog. Outlook, 1978 - 1979 : editorial). In the Damodar Basin in Jharkhand it covers about 8.00% of the total area (Table. 8 - II). It is eating away more areas every year. Unless we study its extent, genesis and the rate at which it is advancing we cannot know its further extension and any conservation planning will be out of context with the physical facts of the situation.

The planning of town settlements, industrial complexes, communication lines, agriculture, etc. require the knowledge of the main relief features and slopes. The distribution of unsuitable forms, e.g. scarps, gorges and terraces, etc. have also to be taken into account.

The above discussions conclude that the aim of geomorphology is to interpret the landforms genetically and also their implications in terrain evalution. Hence, it is an important discipline in view both of its academic and practical aspects.

The landforms that we see in any area are the result of a set of complex forces including subaerial processes, tectonic movements and structural features. The present landforms are a rough indication of the geography of the past and are likely to be influenced by the geography of the future.

By its geology, mineral resources, tectonic history, geomorphic faetures and modern economic importance the Damodar Basin in Jharkhand deserves thorough geomorphic investigation.

The sharply defined 'trough' that the valley is, the steeply inclined stupendous scarps that bound the valley at its head and sides, the rough coincidence of the junction of the Gondwana Sedimentaries and the gneisses with the edges of the valley, the presence of clearly defined parallel faults, the occurrence of a number of cross faults, the existence of a number of igneous intrusives, the relative horizontality of the sedimantaries, the presence of hot springs, the greater steepness of the southern wall of the valley as compared to its northern counterpart, the meandering nature of the main streams and their larger tributaries the relative evenness of the valley floor but-for limited residual rising to the valley of the adjoining plateaus, the four planation surfaces of the catchment bounded by steep scarps, the occurrence of a gentle convex slope towards the upper edge of the scarp, succeeded downwards by a steep rocky free face, succeeded still downwards by a debris slope and still further down by the cancave slopes, the composite nature of hill slopes, the existence of some very impressive waterfalls and rapids, the difficulty in construction of roadways and railways through the scarps, a pronounced degree of soil erosion and precarious agricultural economy, all appear to be related features, whose proper understanding and correlation must rest with a geomorphic study of the area. The present study has attempted to elucidate and explain above - noted features.

1.2. Delimitation of the Area :

Lying between 23^0 28′32" N.Latitudes and 24^0 31' 30" N.Latitudes, 84^0 41′00" E. Longitudes and 86^0 51′45" E. Longitudes, the Damodar Basin in Jharkhand (wherever the term Basin is used capital 'B' it refers to the Damodar Basin in Jharkhand), India (Figs,1A, 1B & 1E) is one of the important geomorphic units of Chotangpur Highlands (Fig.1C). This is an 'exorheic basin' (Fairbridge, 1968 : 284) which debouches into the Hooghly and finally to the Bay of Bengal.

The Damodar is a drainage Basin which includes its whole catchment area. Thus, 'the concept of catchment makes it imperative to include all the affuent streams which are tributary to the main stream holding the basal position in the drainage taxonomy of the valley ' (Raza, M. et al. 1978 :11).

The Damodar is the main stream. Numerous streams join the Damodar from the north and the south. The Barakar, an important tributary of the Damodar, runs almost parallel to the Damodar before it comes to join the main stream.

The Fig.1-f. shows two distinct sets of drainage basins the Damodar and the Barkar system. These two river basins are sepatated by the high divide projected from the Upper Hazaribag plateau. Both the rivers, the Damodar and the Barkar, flow almost parallel to each other and so are their catchments running longitudinally as more or less narrow strips of land from west to east in case of the Damodar and north - west to south - east in case of the Barakar. The

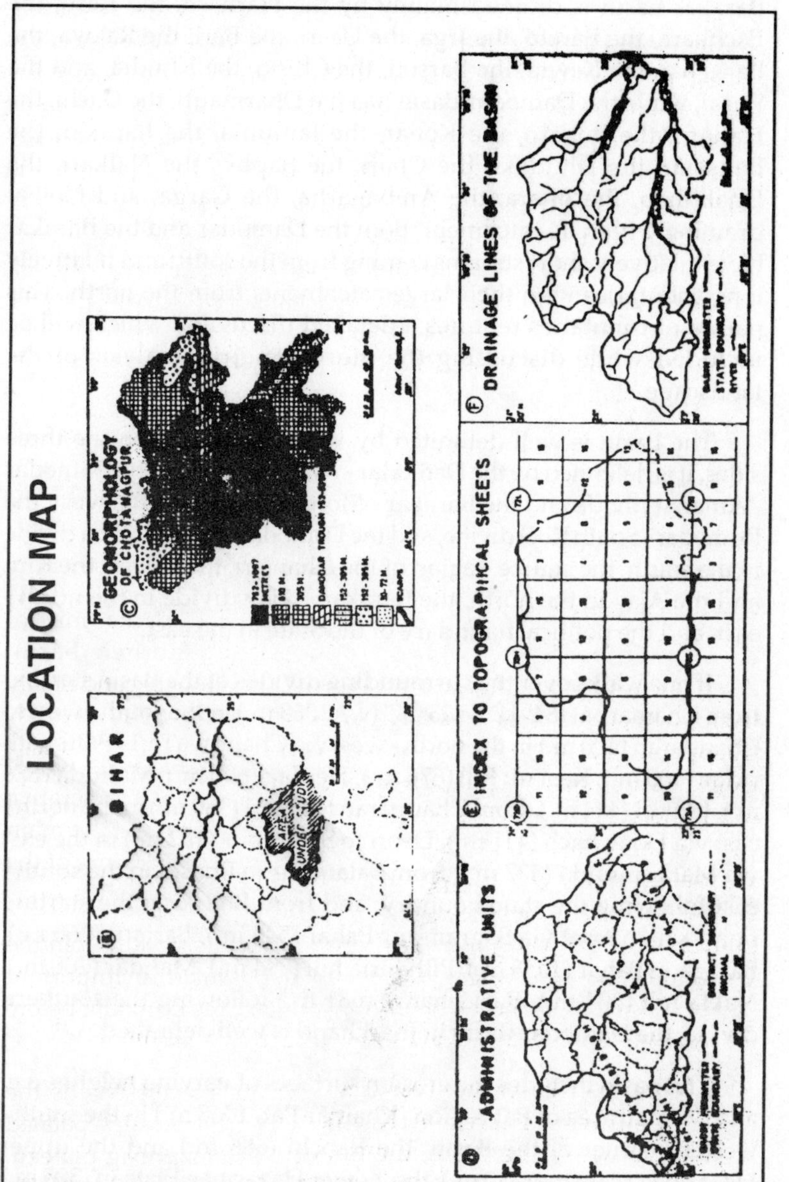
LOCATION MAP
BIHAR
GEOMORPHOLOGY OF CHOTANAGPUR
INDEX TO TOPOGRAPHICAL SHEETS
DRAINAGE LINES OF THE BASIN
ADMINISTRATIVE UNITS

Fig. 1

Barakar basin is drained mainly by the Harnaro, the Keso, the Pachkara, the Bareto, the Irga, the Usari, the Beri, the Rajoya, the Barkijhar, the Kewta, the Barsoti, the Chirki, the Khudia, and the Pusai, while the Damodar Basin has the Dharmauti, the Garhi, the Haharo, the Bokaro, the Konar, the Jamunia, the Barakar, the Bansjhar, the Dhudaki, the Chati, the Saphi , the Nalkari, the Dhabdhab, the Bhera, the Ambagarha, the Garga, and Gowai draining within its catchment. Both the Damodar and the Barakar have relatively short streams coming from the south and relatively long tributaries with their large catchments from the north. This nature of tributaries requires a detailed discussion which will be included while discussing the morphometric analysis of the landscape.

The Basin is well delimited by water - partings on the three sides. It is delimited by the Damodar - Auranga divide, the Damodar - Amanat divide and the Barakar - Mohana divide on the west, the Damodar - South Koel divide, and the Damodar - Subarnrekha divide in the south, the source region of the Dhanjari, the Sakari, the Kiul and the Ajay in the north, the Barakar - Ajay divide in the north - east, and the political boundary of the State in the east.

If one walks over the surrounding divides of the Basin starting from Khamarpat ($84^0$41' E : 23^0 37' N./ 1068 m.) in the south - west to Chauparan (400 m.) in the north - west via ChandwaTori - Balumath ridge (600 m.), Kasiatu hill (676 m.), Hedlag (600 m.), Mahuda (676 m.), Itkhori (444 m.), from Chauparan to Deori (380 m.) in the north - east via Domchach (411 m.), Deori to Salampur (152 m.) in the east via Mahesmunda (327 m.), from Salampur to Bhaski in the south - east following the state boundry; and from Bhaski to the starting point Khamarpat via Sagrampur Pahar (549 m.), Bariatu (360 m.), Baragaon Pahar (1038 m.), Pithauria hill (794 m.), Mandar (700 m.), Nakta hill (804 m.), Barkichanpi (681 m.) following the southern divide, the Damodar Basin in Jharkhand is well delimited.

The Basin includes the erosion surfaces of varying heights, e.g. a very small area of Pat region (Khamar Pat, 1068 m.) in the south - western corner of the Basin, the Ranchi (686 m.) and the upper Hazaribag plateaus (610 m.), the Lower Hazaribag Plateau (305 m.) and the outer Erosion Fringe (152 - 305 m.) in the district of Dhanbad. The well - defined Damodar trough between the Hazaribag and the Ranchi plateaus is of varying width. "The northern and southern limit of the Basin are, therefore, precisely set by the fault - line scraps" (Prasad, 1965 : 57).

In the eastern part of the Basin, the general characteristics are those of an upland country but the relief is subdued and undulations are less pronounced. Consequently areas between ridges are broader and more level, the country is more open and presents the appearance of a series of rolling downs, dotted here and there with isolated residual hills (D.G. Hazaribag, 1917 : 5).

The Damodar landscape has been developed mainly over granites, gneisses of Archaean and sand - stones, and shales of Gondwana formations. The upper part of the Basin has high gradient giving rise to rapid runoff, gully erosion and bank erosion.

The erosion surfaces are relatively flat or gentle sloping, divided into 'dons' and tanrs. Isolated residual hills of resistant rocks break the monotony of the surface. "The alternating dons and tanrs which constitute the universal order of relief on the plateau are actually the water channels and inter - fluvial ridges" (prasad, 1965 : 46).

The Basin covers an area of 16249.66 Km.[2] The Basin extends over *(Fig. 1D)* the entire Dhanbad District, Parts of *anchals* of Jamtara and Narainpur in Santal Parganas, the entire *anchals* of Giridih, Birni, Pirtanr, Dumari, Nawadih, Bermo, Petarbar, Jaridih,Gomia, Bagodar and parts of *anchals* of Gande, Bengabad, Deori, Jamua, and Dhanwar in Giridih District, the entire *anchals* of Gola, Ramgarh, Mandu, Bishnugarh, Barhi, Ichak, Hazaribag, Barkagaon, and parts of anchals of Kodarma, Chauparan and Itkhori in Hazaribag District,parts of *anchals* of chandwa and Balumath in palamu district the entire Khalari and parts of *anchals* of Kuru, Mandar and Ormanjhi in Ranchi District. Thus, the Basin partly or wholly occur in the districts of Hazaribag Giridih, Dhanbad, Ranchi, Palamu and Santal Parganas.

The maximum west - east extension of the Basin is 208 Kms. and maximum north - south extension is 118.4 Kms. The catchment of the Damodar and its tributaries forms a "somewhat kidney - shaped pattern" (Gorrie, 1953 : 3). The most significant point to be noted is that the central part is wider than the eastern and the western parts of the Basin. Thus, the shape and situation of the Basin appeared to be eminently suitable for frequent concentration of run - off in the channels.

Phsiyographically the Basin contains all the four erosion surfaces (Dunn, 1942 : 6 - 14; Ahmad, 1958 : 16). But its discussion is outside the scope of this chapter. However, it will be worth mentioning that the Damodar Basin in Jharkhand provides a museum of topographic features which attract any person of interst for geomorphological study to unravel the past, analyse the present denudation and to predict for the planning of the area under study.

1.3. Previous Study :

The geomorphological study of the Damodar Basin in Jharkhand, as a whole, was not made previous to this work. The diagnostic survey of the Basin jointly carried out by the Calcutta University, the Patna University and Indain Institute of Technology, Kharagpur, relates generally to the human aspects. Several books, theses and research articles have been published on the Geomorphology of the highlands of Chotanagpur and their regions. The Basin under study is a part thereof.

The geomorphological studies pertaining to the highlands of Chotanagpur, in general, have been initiated by geologists and advanced by geographers. Detailed geological pieces of information have been recorded by eminent geologist notably Oldham (1893), Pascoe (1950), Krishnan (1953, 1956), Wadia (1975), Fox (1930, 1931, 1934), and Gee (1932). Krishanan had discussed the structureand tectonic history of Peninsular India. F. Ahmad (1960 - 1961) attempted to contradict the 'rift valley' concept of Fox and others and put forward a 'goesynclinal concept' to explain the 'Gondwana trough' in Peninsular India.

At first, Dunn (1939) interpreted the evolution of the four erosion surfaces in Chotanagpur plateau (including the area under study). He supposed that uplifts were the root cause of the different erosion surfaces and supported his views in his next work in 1941 : 42. He is of opinion that the different erosion surfaces of the area are peneplains uplifted during the Himalayan movements.

S.C. Chatterji (1945) discussed some aspects of geomorphology of Ranchi Plateau. He (1946) has traced the physiographic evolution of Chotanagpur plateau. He contradicted the uplift theory of Dunn and proposed a 'differntial erosion' theory for geomorphic interpretation of 'Pat Region'.

E. Ahmad (1958) reviewed the earlier works and presented his own views regarding the genesis and evolution of four erosion surfaces of Chotanagpur plateau. His unique paper with two valuable maps contradicted the 'differential erosion' theory of Chatterji and accepted the uplift theory of Dunn to explain the evolution of the four erosion surfaces, i.e. Pat Region, 900 m, Ranchi plateau 610 m., Lower Hazaribag plateau, 305 m. and outer Erosion Fringe, 152 to 305 m. Ahmad (1962) also later gave a detailed treatment of the geomorphology of the peninslur India including Chotanagpur. Ahmad and Devi (1965) traced the evolution of scarps of Chotanagpur Ahmad (1969) recognised the 'Higher Rims'of scraps. He has suggested that these surfaces are the eminences of older tectonic elements. He put forward some points against the geosyncainal concept of F. Ahmad (1960, 61) and supported the 'rift valley' concept of Fox and others for geomorphic interpretation of the Damodar trough. Ahmed (1970) reviewed the erosion surfaces in the proceedings of symposium on 'Erosion Surfaces' of Chotanagpur (Dept. of Geography, R.U. 1968) and also contradicted the views of foreign delegates regarding the geomorphic evolution of Chotanagpur. The foreign geomorphologists envisaged that Chotanagpur was a dead erosion surface and that an earlier pediplain had been cumbered with thick sedimentary cover. Ahmad has discussed this issue in detait in his report and pointed out that this view cannot be accepted unless we believe that all sediments in Chotanagpur are of alluvial origin. This is against all the geological knowledge of the area. Ahmad (1967, 1970) gave detailed and lucid account of the 'gully erosion' in India and assumed continuous uplift of the area as the probable explanation for the genesis and magnitude of gullies. This view is, perhaps, a departure from the climatic or climatogenetic concept of Geomorphology to explain its genesis and magnitude.

Recently, E. Ahmad (1979) presented five papers in the field of Applied Geomorphology, in the Dept. of Geography, Rajasthan University, Jaipur. He boldly discussed the role of geomorphology in the spheres of agriculture, communication and other engineering works.

A. Prasad (1965) presented an excellent paper on the physical landscape of Chotanagpur. The article with one map provides a detailed account of the regional units of Chotanagpur. W. Kirk (1950) presented a detailed account of different aspects of the Damodar valley.

R.P. Singh (1956) was awarded Ph.D. degree by the London University on his work 'Geomorphological Evolution of the Highlands of Chotanagpur and the adjoining Districts in Bihar'. He (1957) traced the stages of structural growth of Chotanagar. He (1958) reviewed structure. Drainage and Morphology of Chotanagpur Highlands. Singh (1970) gave an excellent treatment of topography and towns in the Chotanagar Highlands by tracing the significance of rocks and surface in the distribution of urban settlements.

Sharma S.N. (1968, 1970, 1971) discussed river capture and a general problem in the drainage development in western Chotanagpur. He (1972 - 73) analysed the geomorphic evolution of Singhbum plain.

S.P. Chatterji (1940) described the gneissic topography of Ranchi Plateau. Bagchi and Sengupta (1958) discussed physiographic regions of Ranchi Plateau. Chibbar (1953) discussed the landforms, life and radial drainage of Mount Parasnath. N.K.Bose (1940) gave a note on the topography of Ranchi Plateau. S.C.Bose (1945) gave an account of the landform development in the Giridih - Parasnath Region. He (1953) presented natural divisions of the Damodar. A Sharan (1958) traced the evolution of drainage system in the Damodar Basin. L.N. Ram (1968) determined uplift through profile study in south - east Chotanagpur.

P. Devi (1967) was awarded Ph.D. degree on 'Scarps of Chotanagpur : A study in Physiography and Human Geography'. She (1968, 1970 - 71) discussed the character of slope in the scarp zone of Chotanagpur.

P. Evans (1971) discussed the Tertiary Plateau of Bihar. He attempted a correlation between the plateau surface and the known Tertiary history of the adjacent region of Bengal and Assam. This attempt was facilitated by evidence obtained during the exploration for oil in the Bengal alluvial plain (Mathur, et al. 1964).

Satpathi (1970) presented a broad outline of geomorphology of Singhbhum. He (1972 - 73) made a quantitative analysis of landform with particular reference to the Deo River Basin, Singhbhum. He was awarded D.Lit. degree for his thesis on 'Geomorphology of Singhbhum.' He (1975 - 76) discussed the landscape cycles of Singhbhum. He (1976 - 77) made a unique and systematic study of meander mechanism, geometry of meanders and Sinuosity Indexes of major rivers of Singhbhum.

S. Singh (1977) traced the evolution of Granito - gneissic tors and Cut - Off spur mounds of Ranchi plateau. Betal (1970) identified the slope categories in Damodar valley. Pathak (1970) presented a paper on 'Geomorphology of Chechari Basin. A. Prakash (1976 - 77) discussed the scarps of Jashpur Plateau. He (77 - 78) traced the genesis of laterite. He (1979) was awarded Ph.D. on 'Geomorphology of the Lawa Basin'. V.K.Sharma (1976) discussed some hydrological characters of the Damodar river. Vaidyanadhan and Ramana Rao (1972) discussed correlation and age of erosion surfaces in the iron ore ranges of Bihar and Orissa.

N. Prasad (1971) prepared a Bibliography on Geomorphology of Chotanagpur. He presented a number of papers (1973, 1974, 1977, 1978 - 79, 1979, 1980) which are contribution to the field of geomorphology. He (1979) was awarded D.Lit. on his work on 'Geomorphology of the Barakar Basin.'

Gupta (1979) discussed the extent and genesis of gullies in the Damodar Basin in Bihar. Das (1965), Gupta (1966) and Prasad (1970) submitted their master's theses on the geomorphology of different parts of Chotanagpur.

Thus, we may conclude that the geomorphological study of the Damodar Basin in Jharkhand, as a whole, was not made previous to thiswork. Numerous research workers, however, have attempted to analyse the genesis and evolution of geomorphic features of the highlands of Chotanagpur.

1.4. Sources of Information :

Geomorphological study of the Damodar Basin in Jharkhand is based on varied sources. Primarily basic information about geology, stucture, relief and drainage is gathered from various maps and published records.

Geological information e.g. stratigraphical sequences, structural patterns, etc. has been obtained from the geological maps, records and memories of the G.S.I.

Topographical map may be regarded as a representation or model of the landscape and is a major souce of information for the students of geomorphology. Morphometric analysis e.g. linear, areal and relief aspects of the drainage basin is largely based on topographical sheets.

Recently, the technology of remote sensing has become highly developed, and availibility of aerial photography and imagery has greatly increased. Satellite and high altitude aircraft sensing system now coming into use offer a further range of photographic and imagery materials from which a vast amount of data can be collected (Way, 1973 : 1). These maps and records, however, are treated confidential as the area of study lies in an industrial belt which is taken to be of strategic importance.

Valuable information regarding soil, sediments and hydrology has been received from Soil Conservation Department of D.V.C. Hazaribag.

Some maps and data published in different journals and books have been utilized for geomorphic interpretation of the landscape evolution of the area.

Many geomorphic features which were not clear from the topographical sheets have been verified in the field. It took almost about four months to cover the field in different seasons.

1.5. Methods :

Geomorphological study has become highly technical. Since Davis many new ideas and concepts have been contributed to this field. Methods applied and techniques used have become sophisticated. From mere description of landforms and mapping, the stage has come to study the processes and quantify them.

In the early days of geomorphology two methods were generally applied. They were : descriptive or physiographic and experimental.

Davis (1899) gave a semblance of dynamism to descriptive or physiogrphic method when he proposed the 'cyclic concept' for geomorphic interpretation of the landscape evolution. It is regarded as an example of closed system. Davisian cycle starts with maximum potential energy as a result of maximum height created by the initial uplifts of the area. The height and with it the energy reduces gradually with the progress of the cycle. At last a surface of low relief develops where the energy is the minimum or perhaps the surface is in the true state of equilibrium.

At about the same time as the cyclic model was being developed, denudation chronology came into being, initially quite seperately. "Denudation chronology was based on the notion of historical succession, just as the cycle was founded on the evolutionary concept" (Chorley, 1978 : 5).

Penck (1924) model provides a genetic study of landforms, the geometry of slopes and their relation with structural movements. He regarded surface morphology as the competition between the rate of river downcutting and the rate of tectonic uplift of the area. King (1953, 67) has argued that the world's plain lands are not 'peneplains' (Davis) but pediplains formed by stream incision, slope retreat and pedimentation, and that they consist of coalesced pediments.

Ahmad applied the concept of regional uplifts and parallel retreat of slopes for the geomorphic interpretation of Chotanagpur Plateau (the Damodar Basin in Bihar is part thereof).

The early works have prompted a variety of different methods of approach from which have developed different schools of thought. They are : Climatic, structural and Dynamic.

Climatic Geomorphology studies the evolution of landforms in terms of climate. According to this school of thought climatic parameters directly or indirectly control the genesis of landscape evolution. Climato - genetic geomorphology is a historical analysis of Climatic Geomorphology.

Structural Geomorphology relates morphological changes with structural variations and tectonic movements. This method studies petro - variance, epeiro - variance and base level, etc., in tracing the genesis of the landscape. It is concerned with the possible influences of endogenic processes on the geometry of landforms.

E. Ahmed reviewed the climatogenetic and tectonogenetic models of the landscape evolution and presented his view that climatogenetic geomorphology has minor role to play, the real work is still to be done in the field of the tectonogenetic geomorphology.

Dynamic method is distinct from several other methods e.g.

evolutionary approach, denudation chronology, morphoclimatic and structural methods, etc., for geomorphic interpretation of landforms. This approach is time - independent and appears to be an alternative to the cyclic concept of Davis, which is time - dependent evolutionary method. Dynamic model suggests that landforms are the product of competition between the resistance of crustal materials and forces of denudation (Rice, 1977 : 216). It requires a state of balance between opposing forces. These operate at equal rates and cancel each other to produce a steady state, in which energy is continually entering and leaving the system (Hack, 1960 : 82). According to system theory (Chorley, 1962) it is an example of open system.

Tools of dynamic equilibrium are mathematical models and empirical equations involving the forms and the processes. The main objective of this method is to gather accurate data of measurable features of stream networks, drainage basins and landforms etc. and to put them for geomorphic analysis for the establishment of certain basic laws of stream behaviour and the landscape evolution, etc.

We have noted some of the methods of earlier workers in the foregoing lines. Now we have to mention various methods of approach for the geomorphic interpretation of the landscape of the Damodar Basin in Jharkhand. Geomorphology being a field science aims at understanding the genesis and variety in landscape forms. The several geomorphic forms of Damodar area can hardly be explained and understood with a single methodology.

It is important to note that the subject covers portions of the fields of geology, geography, pedology, meteorology, botany and hydrology, etc. and no one person could possess detailed knowledge of all of these. Thus, field observation of geomorphic forms and processes, etc. to get cetain conclusion regarding geomorphic evolution of the area required a long period and a team of workers. A research worker, perhaps, cannot observe all aspects of geomorphology in detail, who has to complete his scheme within time - bound period. Accessibility is another problem in the way of the field observation. Most of the upper part of the Basin is still densely forested where wild animals may be seen.

In spite of these difficulties emphasis, where ever possible, has been given in the field to observe features of geomorphic significance.

Observations in the field especially included soil profiles, river thalwegs, valley side slopes, scraps, higher rim, waterfalls, pot - holes, river capture, gullies, isolated hills and tors, etc. Influence of geomorphology on communications, settlements and agriculture, etc. was also taken into account. Field measurement of hill slopes and gullies was done. Photographs were taken to illustrate geomorphic units.

Geomorphology of the Damodar Basin could be described on the basis of topographical sheets and field observations. But the author took up some laboratory work in preparing some detailed maps and diagrams. For most of the maps the author preferred one inch topographical sheets. Quarter - Inch map has also been used.

Relief profiles have been constructed to present the vertical aspect of topography in the area. Slope analysis has been done at a great length. Channel gradients have been measured. Composite profiles of stream segments have been constructed following the method used in the book of Strahler. For an average slope map of the Basin the author preferred Wentworth's method.

Maps related to the texture of drainage (spatial and frequency distribution), hydrographic networks, (distribution and density of source heads and confluence points), and relief measures (e.g. relative relief, absolute relief, average relief and dissection index, etc.) have been prepared to study the drainage and morphology of the area. The detailed morphological information regarding drainage composition in terms of stream orders, lengths and areas have been obtained and mapped.

Maps relating to communication and settlement etc. have also been prepared to show the control of geomorphology on human activities.

Perhaps the most fundamental aspects in the geomorphological aspects in the geomorphological make - up of the Damodar Basin in Jharkhand are the planation surfaces and the valley areas. The planation surfaces are mainly composed of old rocks i.e. the granites, gneisess and schists, etc. The trough areas consist of sandstones, shales and coal seams, etc.

These rocks have a great variance in their degree of hardness, coherence and permeability. Their studies are of immense value in understanding how the present landscape has come to be what it is.

The field observations, geological information and maps make it easy to intepret the geomorphological features of the area. They also help in the evaluation of the methods and models of the earlier workers. Modern method studies the forms and the processes. Classical geomorphology explains the genesis with historical approach. The ultimate goal of all these approaches is to interpret morphological variations of the landscape. Change is the rule of the nature. Forms and processes are not an exception. They involve numerous geophysical problems e.g. isostasy, continental drift, tectonic activities, and exogenic forces, etc. The study of any one of them would provide only partial explanation.

Modern methodology is important to give a scientific precision to the landscape evolution. But it studies only the forms and the processes. Morphological variations depend, interalia, on stratigraphical formations and structural patterns, e.g. hardness of rocks, folds, faults and joints in the rocks, etc. Again, tectonic activities are important to note, they include horizontal and vertical movements of the earth. Vertical movements give rise to the surface of the earth to be attacked by the exogenic forces. Thus destructional forces are opposed by the constructional forces. For this we have to go into the geological past.

Dynamic equilibrum explains morphological variations of the Damodar Basin in Jharkhand. But this approach cannot be segragated completely from historical aspects of landform development. Each element of the landscape has evolved through a long period to its present position, and this heritage doubtless influences the processes now acting upon it (Leopold, et al. 1964:8).

For the geomophic analysis of the basin we are concerned with some quantitative techniques. In doing so the author is aware of the fact that use of mathematical model has its own limitations. It also suffers from the limitations of mathematical capability of the research workers, who may not always be well - versed in mathematics. Some research workers doubt the validity of the result because " too many conclusions have been drawn upon inadequate quantitative data we should be careful that geomorphology does not become so immersed in mathematics, physics and chemistry that it ceases to be the study of landforms " (Thornbury, 1954 :8). Thornbury says that a knowledge of mathematics, physics and chemistry never becomes more essential to an understanding of geomorphic discussions than a sound appreciation of lithology, geologic structure, stratigraphy, diastrophic history and climatic influences. Baulig (1950) has stated that the laws of geomorphology are complex, relative and rarely susceptible to numerical expressions.

In spite of these limitations and objections quantitative techniques are being intensively used in the field of geomorphology for obvious reasons and scientific precision. The modern methodology marched from the explanatory descriptive approach of classical geomorphology (Powell, Gilbert, Dutton, Davis, Penck, Fennman, Bailey, Baulig, etc.) to modern methodology of dynamic quantitative geomorphological work of Horton, Schumm, Strahler, and Hack, etc.

The method used is analytical. This method has certain advantages over inductive and deductive methods. Chamberlin's discussion of the methodological value of different approaches is excellent. He described three possible methods. They are the methods of the ruling hypothesis, the method of the working hypothesis, and the method of the multiple working hypothesis.

Like inductive and deductive methods, the dangers of the method of ruling hypothesis are that it considers only one theory. This would base on the more obvious and visible features of the landforms. In this connection it is important to note that only favourable features will be taken into account and features conflicting with the theory would be ignored.

The difference between the method of ruling hypothesis and the method of working hypothesis is that in the first the theory has to be established by the observations, while in the second a working hypothesis is established in order to guide the field work and the collection of data.

The analytical method more or less similar to the method of multiple working hypothesis overcomes the main disadvantages of the other methods.

The author preferred analytical method as it is the safest and most satisfying method of solving geomorphological problem of the Damodar Basin in Jharkhand.

The use of modern methodology does not mean that the author has ignored the historical approach of classical geomorphology.

Quantitative methods have also been applied to obtain the scientific precision. At the same time he has taken account of the historical method too, to preserve the virtues of both the methods. It has given an analytical and explanatory tint to the study. The treatment thus has become impartial as various hypotheses both inductive and deductive have been kept in purview.

2

Geology

Geologically the Damodar Basin in Jharkhand (Fig. 2), at a glance, consists of originally solidified crust of granites and gneisses of Archaean; coal bearing and associated deposits of Gondwanas; traps and inter - traps of Mesozoic and early Tertiary times; deposits of alluvium of Recent age concealing the underlying rocks in the lower parts of the Basin; and of intrusive and exposed rocks intruded in different ages. It is believed that the Damodar Basin along with the rest of the Chotanagpur Highlands has never been submerged en masse or wrinkled into mountain chains since Cambrian (Singh, 1958). It has, however, experienced diastrophism resulting into the Hercynian faults, intrusive rocks and epeirogenic uplift.

2.1. Archaean Formations :

Most of the Damodar Basin in Jharkhand, consisting of crystalline gneisses and granites is supposed to represent an originally solidified crust" (Ahmad, 1965 : 12). The Archaeans occupy about 86 percent of the total area of the Basin. They also form the basement on which Gondwanas were deposited subsequently.

There are two main and distinct groups of the Archaean rocks in the Basin : the granites and gneisses and the Dharwars (the ancient precambrian sedimentary rocks).

It is believed that granties and gneisses are the oldest members of the Archaeans representing the originally solidified crust, now generally known as the ' basement complex' or 'fundamental gneiss' probably, similar to 'peninsular gneiss'. But geological studies in Mysore areas suggest that granites and gneisses are not the oldest members of the Archaeans. They are intrusives in the ancient Precambrian sedimentary rocks (Smeeth, 1901 :19 and Fermor, 1909

: 998). If it is taken to be true then it raises the problem as to what the bases upon which the ancient sedimentary rocks (Dharwars) were deposited are ? Heron (1935) and Radhakrishna (1964 : 33 - 34) assumed that the whole complex of granites and gneisses form the basement on which the lower Dharwars rest.

The table 2 - 1 Clarifies the geological formations of the Basin :

TABLE : 2 - 1
The Geological Informations

Eras		Formations	Period		Earth Movement
		Alluminium	Recent		
	Quart - ernary	Laterite	Pleistocene	Tertiary orogeny	Late Tertiary uplift
		High level Laterite	Pliocene Miocene		Middle Tertiary uplift
Cainozoic		Oligocene			
	Tertiary	Lava Trap	Eocene		Lower Tertiary uplift
		Igneous Intrusions	Cretaceous		(Epeirogenetic)
		Mahadeva Series	Jurassic		
Mesozoic		Panchet Series	Triassic		Hercynian Earth
		Damuda Series	Permian		Movements (Faulted Damodar)
Palaeozoic		Talchir Series	Carboniferous		
		Basic Intrusive, metadolerite, amphibolite and epidiorites,			
		Mica pegmatites and granite pegmatites.			
		Chotanagpur granite gneiss.			Archaean
		Darwara-crushed or fault breccia, garnet-amphibole schists, granulites schists,			Earth Movements.
Archaean	granulites, schists, phyllites, quartzites, etc.				

Source : Singh, 1969 : 6 & 7 and Roy Choudhary, 1957 : 38)

It is important here to note that these rocks (granites gneisses and ancient sedimentary rocks) have been subjected to intense tectonic distrurbances in the subsequent periods with the result that they now appear to be 'interbedded and interfolded' (Pichamuthu, 1971 : 86) and render their correlation uncertain.

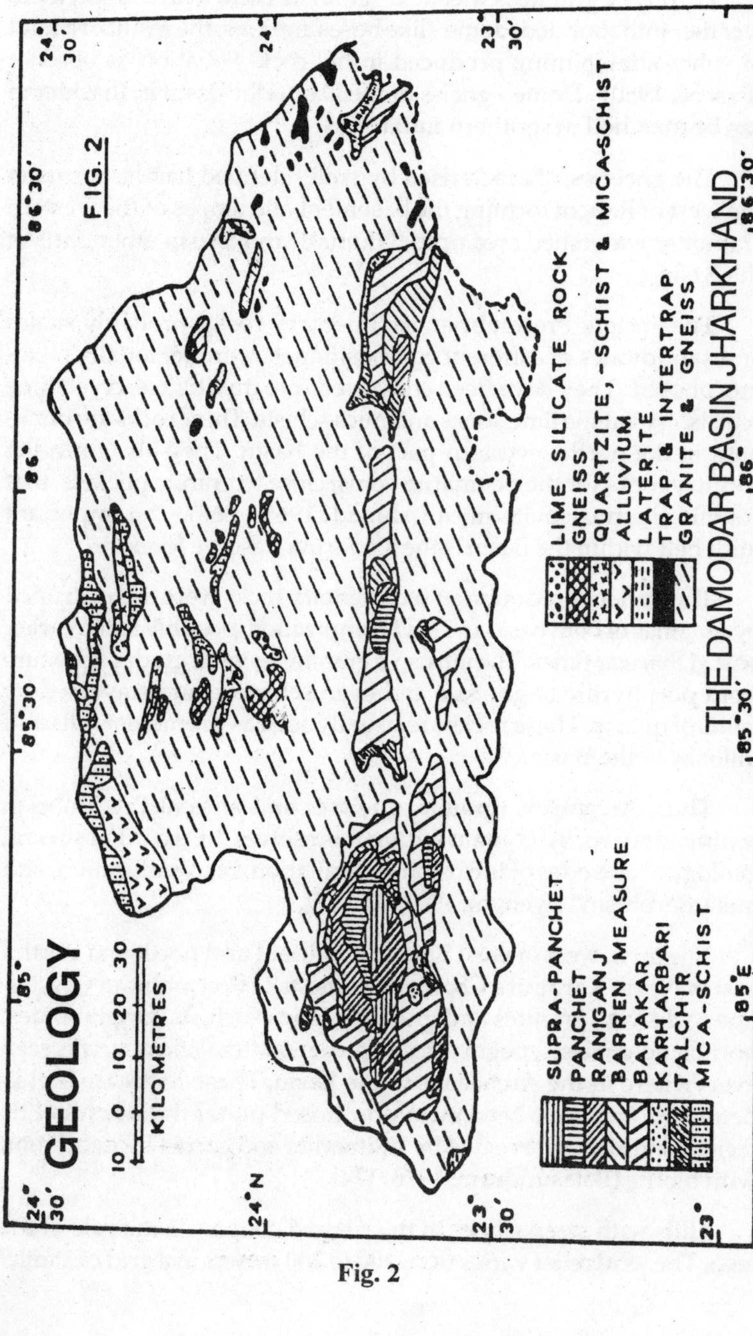
GEOLOGY
FIG. 2
10 0 10 20 30
KILOMETRES
SUPRA-PANCHET
PANCHET
RANIGANJ
BARREN MEASURE
BARAKAR
KARHARBARI
TALCHIR
MICA-SCHIST
LIME SILICATE ROCK
GNEISS
QUARTZITE, -SCHIST & MICA-SCHIST
ALLUVIUM
LATERITE
TRAP & INTERTRAP
GRANITE & GNEISS
THE DAMODAR BASIN JHARKHAND
24° 30'
24° N
23° 30'
23°
85° E
85° 30'
86°
86° 30'

Fig. 2

Porphyric granite is a coarse - grained rock, with a tendency to weather into rounded, dome - like boses and tors, the result of a kind of spheroidal jointing produced inthe rock. Foliation is obscure (Pascoe, 1950) . Dome - gneiss of the Damodar Basin in Jharkhand can be matched in southern India.

The gneisses, characterised by well - defined bands, occurs in the west of Ranchi forming the basement and slopes of the western Damodar watershed area upto Balumath and also in other parts of the Basin.

The ancient Precambrian sedimentary rocks, probably shales and sanddtones of different compositiones, were regionally metamorphosed. They have been converted into quartizites, crystalline schists, crystalline limestones and mica schists. These rocks are dominant in the north - western part of the Basin. They also form the country rocks for the pegmatite veins, close to granite pockets, that contain the muscovite mica (Ahmad, 1965 : 363). The important mica belt within the Basin is the Kodarma Reserve forest.

The ancient precambrian sedimentary rocks were then intruded by magmas of both basic and acid composition, possibly in the earlier post - Dharwar times. The intrusive granites have all grades of texture from porphyritic to gneissic and even schistose and traversed by veins of quartz. These rocks are usually seen as prominent hills and hillocks in the Basin.

Thus, Archaeans (granite, gneisses and ancient Precambrian sedimentary rocks) contain within themselves " a record of several geological episodes of intrusion, assimilation, recrystallisation, and metamorphism" (Iyengar, 1920 : 16 - 17).

The area, west of west Bokaro Coalfield and north and north - east of North Karnpura Coalfield, consists of Precambrian Crystalline rocks viz., granites and gneisses , mica - schists, amphibolites, hornblend gneisses, pegmatites and vien quartz. Vein quartz is seen everywhere in the Archaeans in the Basin. These rocks are highly deformed and have been metamorphosed under the amphibolitic facies condition as revealed by sillimanite and garnet in association with biotite (Balasundram, 1976 :174).

Hills with steep slopes in the rugged country is the rule of the area. The local relief varies from 100 to 200 meters in a grid of 1 mile^2

(2.59 Km.^{2}). The area has drainage density from 3 to 5 miles/ mile^2, stream frequency from 9 to 12 per grid of 1 mile^2 (2.59 Km^2). Source heads 6 to 10 and confluence points from 6 to 9/ mile^2 (2.59 Km.^{2}).

The geological formation near Bagodar, are biotite granite, hornblende gneisses, quartzites and mica schists. These formations have been intruded by amphibolites, epidiorites and meta - dolerites. A few epidote rocks are seen around Barkatha. This geological formation has the maximum development of short steep ridges and isolated hills, which diversify the flat landscape especially in watershed area.

Smaller exposures of east - west trending quartzite, quartz Schists & quartz mica schists occur to the north - east of Hazaribag town flanking the Barakar river. The terrain is rugged. The local relief varies from 25 to 100 meters in a grid of 1 mile^2 (2.59 m^2).

Crystalline limestone, " occuring as narrow conformable bands within the phyllites and calc - schists" (B.D.Gupta, 1972 : 146), is seen in the upper part of the Damodar Basin. It is found at Bundu - Basaria, Lapanga - Bhurkunda, Ray - Bachara, and Babhane - Khelari regions. These beds are striking parallel to and in the close vicinity of the coalfield.

Crystalline limestone strikes east - west and forms a prominent ridge in Khelari area' (Dunn, 1942 : 169). In the area of Hosir - Bachara, the strike is E - W and dips are steeply north. In this region, the limestone forms an east - west series of four hills rising upto 30 to 45 meters.

Schists and calcarious schists are interbedded. Dolerite dykes cut the limestone in Bundu - Basaria area. Two small exposures of limestone are to the north - west of the Konar reservoir.

The lofty Parasnath hill (1366 meters), standing abruptly above the surrounding country, is an 'igneous intrusive body' — Ahmad, 1965 : 26). It consists of hard compact granite similar to the ' pyroxene granite of the Nilgiri Hills' (D.G. Manbhum, 1911 : 28).

Chibber (1953) identified various rocks in the Parasnath hills. Eclogite, gabro, diorite granite, gneiss and quartzites are some of them.

The bold relief (highest in the Basin), low to medium drainage texture and hydrographic network of low to medium density are observed. The most significant point is that some of the streams have lost their existence at the foot of the hills.

Broad dykes and masses of epidiorites intruding the granites and gneisses are abundant in some localities. They are well seen on the road between Dhanbad and Topchanchi.' They are in all probability of the same age as the newer dolerite of the Singhbhum area' (Pascoe, 1950 : 236).

The rocks of the area east of Parasnath are pelitic and psammo pelitic meta - sediments of Archaean age, isoclinally folded with contemporaneous gneisses. The original pelitic metasediments have been regionally metamorphosed. The epidioritic sill which is folded with the older rocks. occurs as isolated (Sinha and Dave, 1963 : 1145).

The above analysis conclude that Archaean formations have a variety of rocks (Table 2 - I) giving rise to a variety of landscape. The Archaeans have apparently, developed two types of landforms, firstly, an undulating and rolling surface (planation surface) with moderate to low local relief, medium drainage texture and medium hydrographic network, and secondly, hilly surface with steep slopes (scarps), thin soil cover, high local relief, low drainage texture and low hydrographic network, all these mainly due to the rocks which are resistant to erosion. So, there is impact of Archaeans on morphological features. The morphometric analysis of drainage, relief and slope in the subsequent chapters makes all these clear.

2.2. Gondwana Formation :

The Gondwana formations in the Damodar Basin in Jharkhand represent rocks of the late Palaeozoic and Mesozoic eras. The subsequent geological history of the Basin reveals that parts of the Archaean basement subsided as a result of block - faulting and formed a deep trough which was filled up later by river - borne deposits, decayed tree trunks and other woody materials during the Permo - Carboniferous and Mesozoic times which are grouped together under the Gondwana System.

Without going into any dispute whether the Gondwanas were deposited in ' down faulted rift valleys' (Fox, 1931 : 241; Wadia,

1975 : 165; Krishnan, 1956 : 266 ; E. Ahmad, 1969; 1970, Prasad, 1979 & Gupta, 1966) or ' geosynclinal troughs (F.Ahmad, 1960; 1961), we are concerned here only with the morphological features developed thereupon.

The major basins of the Gondawana rocks in the Damodar Basin run in an east - west direction, parallel to the general structural trend of the Archaeans conforming approximately to the direction of the Auranga and the Damodar valleys from Hutar coalfield in the west to the Jharia and Raniganj coalfields in the east. Giridih and a few other pockets lie outside the main line of strike.

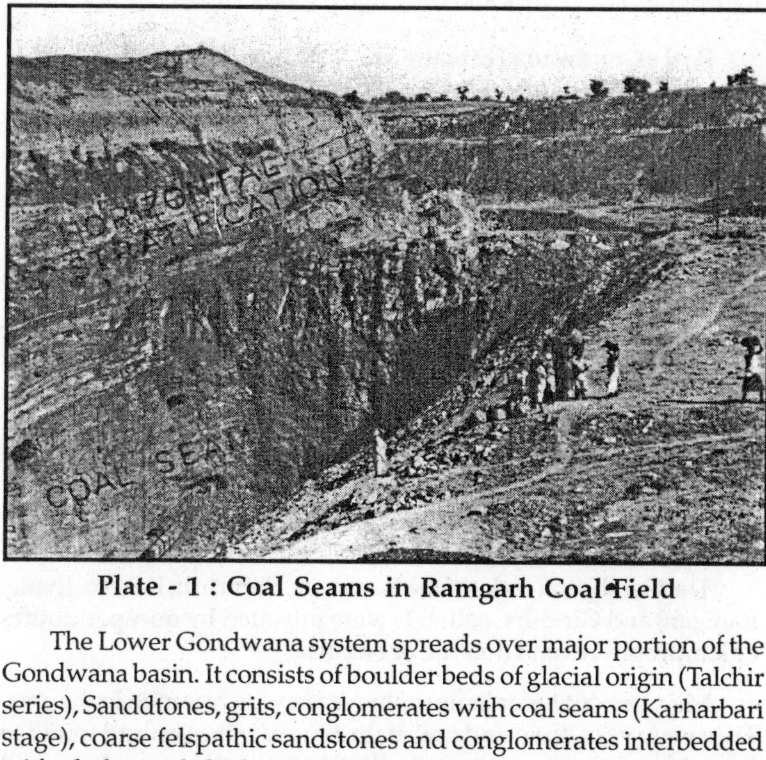

Plate - 1 : Coal Seams in Ramgarh Coal Field

The Lower Gondwana system spreads over major portion of the Gondwana basin. It consists of boulder beds of glacial origin (Talchir series), Sanddtones, grits, conglomerates with coal seams (Karharbari stage), coarse felspathic sandstones and conglomerates interbedded with shales and thick seams of coal (Barakar stage), shales with clay, ironstone nodules devoid of coal (Barren measures), fine sandstone with extensives deposites of coal seams in Raniganj coal fields (Raniganj stage), and fine red clays and sandstones (Panchet series). These formations belong to the Permian and Triassic.

The Upper Gondwanas (consisting of Mahadevas, Rajmahal, Jabalpur and Umia series) are only represented in the area by the supera - Panchet series. They occur mainly in the Karnpura and the Bokaro coalfields forming summits and upper steep slopes of the Morcha Pahar & Lugu hill respectively. The rocks are ferruginous sandstones and are very resistant to weathering.

Thus there is variety of Gondwana formation within the Gondwana basin itself, and consequently there is variety of morphological features in the coal - bearing basins too. The variety is also observed in the characteristics of slopes, drainage texture, relief, hydrographic network and drainage pattern, etc.

2.3. Post - Gondwana Formations :

The post - Gondwana formations of the geomorphological importance in the Damodar Basin in Jharkhand include mainly the igneous activities in both the forms, lava trap and instrusions.

2.3.1. Lava - Trap :

The south - western corner of the Basin experienced lava outpour, probably, during the Cretaceous - Eocene period. This trap is actually a continuation of Deccan Trap. Most of the trap has been converted into high level laterite.

The summit of the hill (Khamar pat) is flat. It slopes steeply the average gradient of which is more than 8^0 and local relief is more than 400 meters.

2.3.2. Igneous Intrusives :

The Gondwana rocks of the Karnpura, Ramgarh, Bokaro, Jharia, Raniganj and Giridih Coalfields were intruded by micaperidotites or lamprophyres and dolerite or basalts.

Mica peridotites or lamprophyres are very common in the area. They were very 'fluid and hot' at the time of intrusion and found to favour weak horizons, ' generally the fault planes, and top and bottom of the coal seams which are the paths of minimum resistence' (Dutta, 1971 : 77 - 78).

Unlike the dolerites, mica peridotites are ' very irregular in shape often branching out and even anastomosing within the strata

(Chatterji, et al, 1970 - :7). Instead of forming sheet like sills and caking a few meters of a coal seam in a regular manner, many of the intrusions are in the form of arterial or vein - like sills and do not always keep to one horizon in a coal seam. The result is that the whole seam is rendered unattractive by these, anastomising, root - like occurrences of peridotite - each with its seath of caked coal (Fox, 1931 : 60).

Dolerite dykes often follow straight courses without any reference to local structure and faults. These dykes may have some genetic connection with the Deccan Trap lavas found in Netarhat and Khamarpat area. (Fox, 1930 :22). Lava trap exposures of Cretaceous - Eocene periods extend as far as Lohardaga in the Ranchi District and Khamarpat area in the south - western corner of the Basin. If the Deccan traps do extend so far out, it is also reasonable to assume that dykes connected with them would be found at least in the Karanpura coalfield of the Upper Damodar Basin. These dykes are characterised by their great thickness, typical jointing and spheroidal weathering. The trend across the strata is regular, quite unattracted by the coal seams and almost uninfluenced by fault planes.

As regards their ages, mica - peridotites and dolerites belong to two different ages. In the Giridih coalfield, " a large dyke of basalt cuts a thin dyke of peridotite' (Holland, 1895 : 130). It may be concluded that mica - peridotite is earlier and dolerite dykes are later in age.

2.4. Recent Formations :

Most of the area of the Damodar Basin in Jharkhand is covered with soil and sediments of Recent formations. They include the residual soils mostly derived from the weathering of bed rocks or developed in situ and alluvial sands and silts of rivers. The alluviums made up of silts, clays and fine sands of Recent periods are found in the lower portion of the Basin. These deposites occur on the bank of the Damodar, the Jamunia, the Ijri, the Gowai, the Barakar and the Khudia rivers.

The alluvium on the bank of the Ijri river near Bhojudih is reddish sandy clay. A small area of reddish sandy clay occurs on the eastern bank of the Jamunia river. All other river silts are distinctly sandy.

In addition to recent alluvium, patches of older alluviums are also seen in the area under reference. The situation of older alluvium of Amlabad, Swardih, Sudamdih and Tasara suggests that the Damodar river of then flowed southwards from both of Amlabad to near Karadih, then turning sharply north - east towards Swardih was deflected eastwards north of Sudamdih but south of the site of Chasnala colliery towards Hetkendra; then it changed its course to south east after crossing the site of Domohani. At a later day it appears to have shortened its bend south of Bhaunra and to have made an island of Sudamdih before adopting the channel it now occupies from above Amlabad (Fox, 1930 : 106).

In the vicinity of the Barakar and the Damodar rivers, patches of gravels with occasional sands occur, in some cases well above the present high flood level of the rivers. These gravels are frequently highly Kunkeriferous (Gee, 1932 : 78). Massive beds of 'Kunkers' are seen adjoining the Barakar river near Hatinal south of Chirkunda.

The discussion on the geology of the Damodar Basin in Jharkhand concludes that the Basin is formed of rocks of differnt geological ages. It is true that the major portion of the Basin (about 86% of the total area) is composed of the Archaeans. Yet other formations are also important. This is why the morphological features vary in magnitude and dimensions. Steep and bare rocky scarps, rugged topography, long, deep and terraced river valleys, waterfalls and rapids, residual hills and many other features are the result of this geological variance and epeirogenetic uplift of the area during the Himalayan orogeny.

3

Structural Characteristics and Tectonic History

Geological structure has a major role in the morphological variations of the landscape of the Damodar Basin in Jharkhand. The processes of denudation etch out unresistant rocks and lines of weakness into valleys and lowlands and leave the more resistant strata standing as uplands. hills and ridges, etc.

The objective here is to illustrate with the aid of examples taken from various parts of the Damodar Basin in Jharkhand, how the geological structure controls the development of landforms. It is to be remembered that landforms cannot be ascribed to one cause. They are, instead, the result of a complex interplay between several physical factors and processes, both endogenetic and exogenetic.

The term structural means not only the attitude of rocks but it also include faults, folds, the bedding planes, the physical hardness and softness of rocks as well as the regional relationship of the rock - strata (Thornbury, 1954 : 18).

In general, the structural features of rocks are much older than geomorphic forms developed upon them. Such major structural features as folds and faults may go back to far distant periods of diastrophism. Even in areas of as recent diastrophism as that of the Pleistocene, it is difficult to find uneroded folds. Hence, as a general principle, we may assume that most rock structures were established long before the landforms which exist upon them (Thornbury, 1954 : 18).

The structural history of the Damodar in Jharkhand as well as the Chotanagpur Plateau and Peninsular India is closely linked up with the earlier supercontinent of Gondwanaland which, probably, disrupted in the Mid - Mesozoic time (Singh, 1969 : 13). Being composed of the older Archaean rocks, fundamentally, the Damodar Basin, in particualr, did not undergo horizontal movement. But it could be free from the earth movements which followed in the geological history. The Hercynian revolutions resulted in long linear fractures (faults), a series of fault - troughs. The Gondwanas were deposited in these block - faulted rift valleys and these valleys resulted in coal bearing basins. During the Tertiary times, the Basin along with the other parts of the Indian Peninsulas, formed a foreland which gave rise to the Himalayas, and the Basin itself was uplifted epeirogenetically. The margins of the Permo - Carboniferous down - faulted trough were uplifted which presently form steep scarps on either side of the trough. This uplift has also been responsible for initiating a new denudational cycle in the Basin.

So, the Basin bears scars of all these events of structural history. The igneous intrusions throughout the ages, however, complicated the matter.

We can, therefore, present structural characteristics of the Basin, in brief, specially in relation to the morphological features guided and influenced thereby.

3.1. Precambrian Structure :

About 86 per cent of the total area of the Basin is composed of Archaean rocks, viz. granites, gneisses, quartz micaschists, hornblende gneisses, etc. The Dharwars formed the cover of the batholiths. Structural trend lines of the Archaean rocks open the clue that the region was folded in the east - west extension. The denuded surface of the area reveal the fact that "the region must have been subject to a continual series of tectonic movements forming complex fold structures (Singh, 1957 : 86). Contortions, overfolds and recumbent folds are common in the Archaean rocks. Dips generally vary from 20^0 to 70^0. The main folding took place during the Dharwarian times. Long continued denudation has exposed the formations, e.g. in the Chutupalu area and the Kodarma area,etc. They reveal 'the very faberic of the structural story' of the area but the remains of Precambrian formations with their complex structure render their correlation uncertain.

Rocks of Precambrian formation vary widely in their resistance to denudation. The agents of denudation etch them out according to their degree of resistance. Gneisses are strong and resistant, which like granite, generally, form bold topography.

Some of the hills in the area consist of huge dome - shaped masses of granite gneisses. The rocks of these hills have been called the 'dome gneisses owing to their perticular weathering into dome - like hummocks and ellipsoidal masses due to exfoliation. These domes from a very striking feature of the landscape along the northern fringe of the Basin.

Plate - 2 : Pona Tongri : The Process of exfoliation is visible

Although exfoliation and onion - like weathering is a hack-neyed expression in geomorphology, probably, non has explained this phenomenon in clear terms except Ahmad who says that "the originally rectangular blocks resulting from the prominent joints, which probably developed as the magma solidified and contracted, have been rounded by exfoliation. At the corners of a ractangular solid block three exposed surfaces meet. Hence the disintegration force of weathering agencies is roughly trebled. At the junction of two faces is double that on any single face of the block. Hence, a ractangular solid is reduced to a rounded one in due course. Later onion like peeling off is parallel to the rounded exterior of the mass" (Ahmad, 1976 : 51).

The rocks of the Archaean formation possess a constant, more or less foliated of banded structure designated as 'Gneissic'. The gneisses, frequently shows great lack of uniformity either to composition or of structure and varies from place to place. At times it is very finely foliated, with foliation of extreme thinness e.g. near Ichak and Padma, etc. At other times, there is hardly any foliation or schistosity at all, the rock looking perfectly granitoid in appearance as is observed particularly in Jiramo (near Atka) and Ghiahi (near Padma) hills (Prasad, 1979 : 82).

The way in which the Archaean crystalline rock was banded is not well understood yet, and "various modes of formation have been ascribed to these rocks" (Wadia, 1975 : 73).

In the Damodar Basin in Jharkhand,"though the banding may be original due to tectonic forces during the intrusion of the granite magma, there is no doubt that the rock has suffered from later regional metamorphism" (Roy Choudhury, 1957 : 38).

The vast area of the Basin is a series of plantation surfaces entirely composed of Archaeans. These rolling surface dominated by dons and tanrs indicate that the area has been uplifted epeirogenetically in the recent past.

Schists and quartzites in the northern part of the Basin form hill and uplands. Mahabar hill is composed of quartzites. Here the local relief is about 250 metres. The drainage texture is coarse and density of source heads and confluence points is low.

The rocks of Nurango area (24^0 10'N : 86^0 O5'E) are mica - schists, quartzites, phyllites and biotite gneiss of the Archaean age. "The schists are folded into a series of anticlines and synclines and later along with the gneisses are involved in large - scale faulting. These are traversed by acid and basic dykes and numerous quartz veins of varying dimensions" (Murthy, 1970 : 41). Being more resistant to erosion they form conspicuous topography. Local relief varies from 50 to 100 metres.

Diorite rocks which generally rise as small hills from the crystalline basement rocks may be observed in the Giridih area. They also occur as dykes in various places in the Basin parallel to the foliation of the pre - existing rocks.

Examples of quartzites and mica - schists may be seen near Gajadhar tanr about 10 km. north of Dumari. These rocks, resistant to denudation form bold relief i.e., hills in the area.

The Parasnath hill is an igneous intrusive body. It is composed of hard compact granite. The bold relief (highest in the Basin) and radial drainage pattern may be observed.

In the areas north - east and north of North Karnpura coalfield and west of Bokaro coalfield, the rock types structurally appear to have been regionally folded into isoclinal type of folds with moderate plunge towards north - east, the axial plane being overturned to the south. The superposition of a later phase of folds movement has folded the earlier folds into open type of folds with moderately steep plunge towards north. The regional strike in the area is ESE - WNW with variable amounts of dip eiher to north or south.

North of Dighi Basti ($24^0 08'$N : $85^0 32'$E) gradations from mica - schists to quartz sericite mica - schists, quartziteschists and quarzites may be observed. "Local puckering and foldings are common in gneisses and schists' (Krishanan, 1958 : 34). Local relief is about 150 metres.

Plate - 3 : A Hill worn down to heap of Boulders

Joints in rocks have a great significance in the development of landforms. They strongly effect the 'embroidery of the relief' (Sparks, 1971 : 107). several hills which stretch from Birni to Giridih and also in other parts of the Basin, have been converted into blocks of weathered rocks. Some of the hills have been reduced to a mere heap of boulders most probably, due to joints in rocks.

Pattern of joints in rocks controls the development of drainage lines in the Basin. The west - east joints have facilitated the rivers to carve out deeper channels with east - west alignment leaving steep - sided walls on either side. The Barakar, the Konar, the Barsoti and the Siwani rivers follow the primary and major set of joints from west to east. Tributary streams seem to follow the secondary and minor joint systems in the Basin.

It is also to be noted that where the rivers flow along the joints they cut the valleys deep with steep - sided banks , as are observed at several places like Chandro and Cherhi etc. At the same time, the joints have another influence also on morphology. Where the rivers flow across the joints rapid or water falls are generally seen developed upon this sructure.

3.2. Gondwana Structure :

The Gondwana basins are peculier in their distribution of geometric configuration and structural set up. The major basins are associated with river valleys delineated along certain lineally arranged fault zones. The magnitude of the faults on the two sides is very unequal. This has the effect of producing a dip of strata towards the more faulted side with the greater throw.

The southern side of the Basin is marked by normal gravity fault, known as the 'boundary fault'. The sedimentary beds show ' a homoclinal dip towards the 'fault Zone'. The dip of the strata becomes steep near the boundary faults in the oppsite direction and , therefore, " an asymmetrical synclinal structure is observed". Gross faults have affected the normal deposition to a great extent. These faults always affect both the Gondwana and the basement rocks and usually 'strike slip' in nature.

Besides, an intricate network of intra - basinal faults affects the sedimentary formations of the Basin. These faults are normal gravity type but some minor ' reverse faults' are also observed in the Basin.

The Jharia coalfield area is an important example. These faults show varying displacement both in strike and dipdirection (Chatterji, 1970).

The thickness of the sedimentary strata and coal seams change abruptly on either side of these faults. Thus, these faults are 'intra - formation' in character.

The Gondwana structures have the following impacts on morphological feature of the Basin.

(i) Morphologically the Basin is a trough area bounded by steep scarps on the three sides - north, sout and west.

(ii) Faulting alone has accounted for the practical horizontality of the sedimentary beds in the trough area and these beds have different drainage and relief aspects than those on the Archaeans.

(iii) The control of faults on the development of drainage lines is immense (Fig.13C). The Damodar is itself a fault - guided river. Many tributaries and sub - tributaries of the Damodar are guided by the faults.

(iv) Faulting provides a link between the surface and deep - seated area. This is suitable site for the hot springs. Bongahara hot spring in the Chotha basin near Kuju and Rajrappa hot spring at Rajrappa near Gola are some of the notable examples .

(v) The Gondwana trough, through which the Damodar flows, appears to be more hilly and undulating than the adjacent planation surfaces, as a result of differential erosion of rocks of varying hardness.

(vi) The principal basins of Gondwana rocks are distributed in an east - west belt, parallel to the general structural trend of the Archaeans and conforming approximately to the direction of the Auranga and the Damodar Valleys from Hutar Coalfield in the west to the Jharia and Raniganj Coalfield in the east.

(vii) The Gondwana strata are mainly composed of sandstones and shales with coal seams. They give rise to topographic irregularities. The sandstones generally form bold topography like ridges and hills and shales form the valleys

The coarse - grained sandstones, barren of carbonaceous horizons, being hard, and massive and resistant to erosion, are exposed as prominent ridges and hills.

It is usually Panchet and Mahadevan Sandstones which are particularly hard and form conspicuous hills and ridges. The Satpahari hill, the Mahudi hill and the Morcha Pahar in the central part of the Karnpura coalfield and the Lugu hill in the Bokaro coalfield are notable examples. These hills dominate the landscape of their respective areas. Similarly the Kathara Pahar in the eastern portion of Kathara - Jhirki area in the Bokaro coalfield is a prominent ridge formed entirely of massive sandstones. Again, Bandh Pahar and several other hills, lying in the southern portion of Kathara-Jhirki area, stand as high hills due to differential erosion. These hills and ridges are, however, not continuous along the southern periphery due to "interruption by oblique faults" (Datta, et al, 1963: 102).

The shales generally form lowlands. But when they have been burnt due to spontaneous comburtion (?) of the underlying coal seam, they have become hard and resistant to denudation as a result of which they form hillock. A bold hillock of indurated shale north of Jarangdih railway station may be observed.

(viii) The hills in Giridih coal field clearly show the dip slopes. Cracking and erosion have proceeded along the joint planes, which are usual along the dip, strike and bedding planes. A huge crack and a slight displacement along it in the direction of the dip may be noticed in a sandstone hill about 4 km. from Giridih (Bose, 1945 : 108).

(ix) The stream beds in Giridih coal - field are cracked in complex patterns and are pitted with pot - holes leaving small pyramids of horizontal beds of shales here and there. The topmost bed in the Talchirs are black carbonaceous shales which generally form low grounds (Bose, 1945 : 108).

3.3. Lava - Structure :

The denuded gneissic surface on the south - western cornor of the Damodar Basin in Jharkhand is capped by horizontal lava trap. The landscape of this region stands out in great contrast to the landscape of the neighbouring areas. The most striking feature of the region is the 'mesalike' flat - topped small plateaus. The present level is on a thick stratum of laterite and not on lava. This leaves the possibility that laterite can behave as a remarkable capping layer in maintaining a horizontal surface. This may follow not only the great resistance offered by hard ferruginous granules but also undoubtedly because water is easily passed down through the wide spaces and has less marked erosion on the surface (Ahmad, 1970 : 11).

3.4. Tertiary Epeirogeny :

The work of denudation proceeded unwaited before the Tertiary epeirogeny and consequently the entire surface was assumed to be reduced more or less to the base level. The surfaces were known as Pre-Tertiary Peneplains or early Cainozoic planation surfaces. During the Teritary times epeirogenetic events seem to have expressed in the form of warps, uplifts and even tilts. These events have influenced the Basin, in particular in the following manners:-

(i) The Tertiary cycle was initiated with the epeirogenetic uplift during the first Himalayan Orogeny. The cycle was interrupted by the second and the last Himalayan earth movements during Mid-Miocene and Pleistocene periods respectively.

(ii) A series of planation surfaces, e.g. the Khamar Pat, the Ranchi & the Upper Hazaribag plateaus, the lower Hazaribag plateau and the Dhanbad plateau are the outcome of the successive and intermittent uplift of the area.

(iii) Epeirogenetic uplifts initiated a new denudational cycle which has carved out the present landscape. They also explain the interruptions in the cycle of erosion.

(iv) The well - marked Damodar trough running west - east through Chotanagpur Plateau and bounded by steep scarps on both the sides is an uplifted and re - excavated feature. The epeirogenetic uplift has 'led to the re - excavation of the softer belt of Gondwana sediments" (Ahmad, 1969:VIII) on the same site and has unwrapped again the steep - sided scarps, presently known as 'fault - line scarps'.

(v) Many features of geomorphic significance, e.g., waterfalls, rapids, valley - in - valley, the magnitude of gully, and incised meanders, etc. are the result of epeirogenetic uplift of the rea. "The uplift has provided the general physical basis of prominent down - cutting and tumbling down of the flowing water during rains at the edges of stream banks (Ahmad, 1969:VI). In the Damodar Basin in Jharkhand numerous gullies and cirque - like incipient gullies are explained also by the uplift of the area. These form at places miniature landforms.

(vi) Most of the hills, hillocks, and summit plains, etc. represent the remnant parts of the old surfaces. They are partly the result of the uplift of the area which have provided new denudational cycle.

4

Physiography

The Damodar Basin in Jharkhand is a region of great physical inequalities. It is an old polycyclic area, which has been peneplained and rejuvenated several times. The present landscape in the Basin is the outcome of the latest phase of the physical history and subsequent subaerial denudation upon different structural patterns.

A close examination of the existing contours i.e. the actual landforms (Fig-3) shows that on different planation surfaces isolated hills, ridges, river valleys and spurs add to the variety of topographical features and bring about strong contrasts that justify the following physiographic divisions of the area.

(1) The Khamar Pat Region.
(2) The Ranchi and the Upper Hazaribag planation surfaces.
(3) The Interplateau Ridge.
(4) The Kodarma - Giridih plateau.
(5) The Outer Erosion Fringe.
(6) The Damodar Valley.

4.1. The Khamarpat Region :

The landscape of the Khamar Pat region stands out in great contrast to the landscape of the neighbouring areas. It is an extension of the main Pat Region. It has a flat top at an elevation of about 1067.5 meters. It owes its existence to "the basaltic cover which is essentially horizontal in attitude" (Singh, 1969 : 24). The present level is on a thick stratum of laterite and not on lava. This leaves the possibility that laterite can behave as a remarkable capping layer in maintaining a horizontal surface. This may follow not only from the great resistance offered by hard ferruginous granules but also undoubtedly because water is easily passed down through the wide pore spaces and the less marked erosion on the surface (Ahmad, 1970 :11).

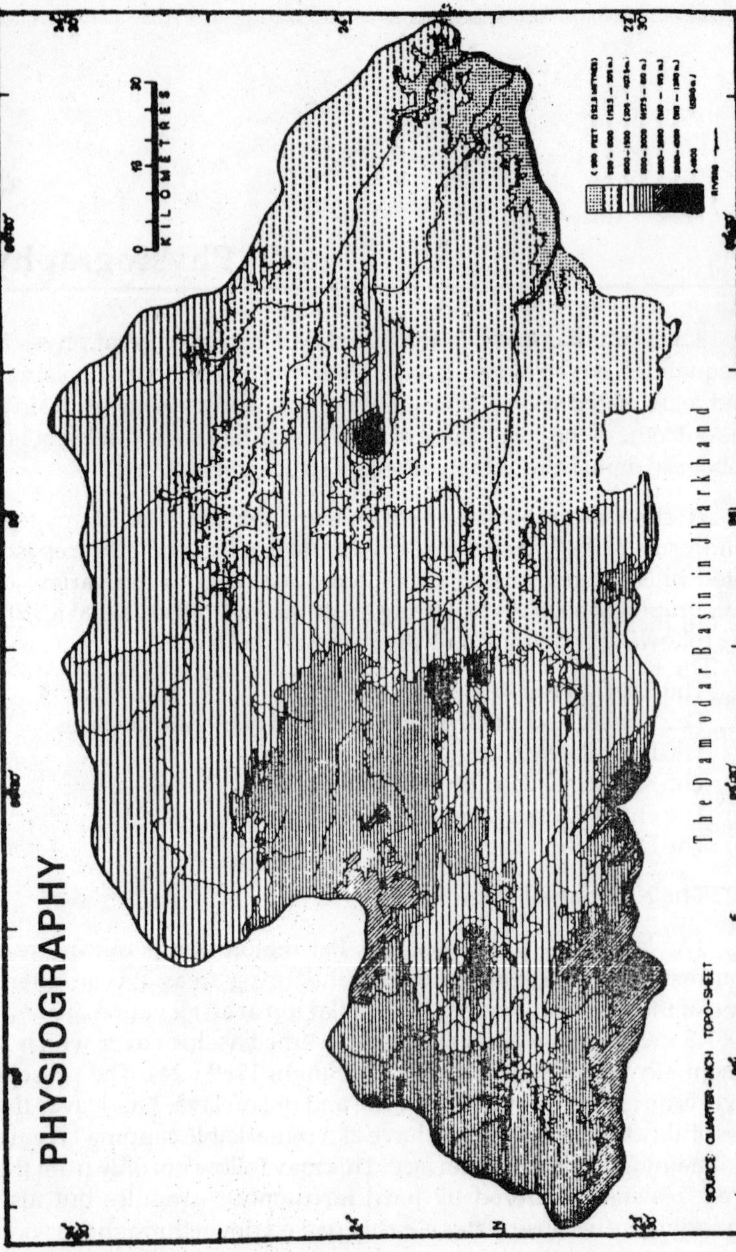
PHYSIOGRAPHY
0 15 30
KILOMETRES
The Damodar Basin in Jharkhand
SOURCE: QUARTER-INCH TOPO-SHEET

Fig. 3

The Khamar Pat is a part of the oldest planation surface of the area. It overlooks the Damodar Basin with steep scarps. The average slope is more than 8^0 (Fig. 15), but it rises upto 55^0 or even more. The local relief on the flat surface is less than 2 meters. But if we take into account the upper surface (1068 m.) and the Deonad basin (488 m.), the local relief is 580 meters. The Birjangha Pahar (970 m.) is separated from the Khamar Pat (1068 m.) by a gap (600 m.). Numerous tributaries of the Damodar, the Auranga and the south Koel originate from the Khamar Pat area. They radiate to form the radial pattern mainly guided by the steep slopes.

4.2. The Ranchi and the Upper Hazaribag Planation Surfaces :

The contour of 600 m. is significent in the Damodar Basin in Jharkhand. It represents the Ranchi and the Upper Hazaribag planation surfaces of granites and gneisses. That it is an uplifted peneplain "is evident from the uniformity of the plateau surface, isolated residual low hills, high degree of truncation of structure, the rivers being indifferent to softer Dharwars and harder gneiss and granites" (Ahmad, 1976 :27).

4.2.1. Ranchi Planation Surface :

The Ranchi planation surface is practically level with undulations of highlands and lowlands locally known as Tanrs and Dons respectively. The general level of the plateau varies from 600m. to 700 m. (Amsl). The average slope of the plateau surface is less than 3^0 (Fig. 15). But it increases towards the edge of the plateau where it overlooks the Damodar trough with steep scarps. The relative relief (Fig. 17) is less than 50 metres in a grid of a $mile^2$ (2.59 km.sq.). But it increases upto 100 meters or even more near the hills and ridges. In the scarp zone the relative relief is high. It is the highest (516 m.) near Baragaon Pahar.These hills break the monotony of the plateau surface. Nakta (23^0 32' N 84^0 57' E.:805 m.) and Silsikhia (23^0 34' N. 84^0 50' E. : 671 m.) are some of the important hills. The landscape near pithouria is dominated by the hills of weathered blocks. The control of joints on the evolution of landscape may clearly be observed here.

Numerous streams originate from the plateau surface. The long profiles of the streams have been broken by the scarp. The rapid flow of water in the scarp zone has dissected the edge of the plateau. The Saphi, the Nalkari, the Dhabdhab and the Bhera are some of the important streams which meet the Damodar from the south.

4.2.2.The Upper Hazaribag Planation Surface :

In the north of the Damodar trough the area of 600 metres level lies in the Upper Hazaribag planantion surface. The Damodar trough separates it from the Ranchi Planation surface. The isolated hills, the ridges and the river valleys break the monotony of the surface. Over the undulating surface of the uplifted peneplain project a number of peaks and hills varying in elevation and form (Singh, 1969 :31). The Chandwar Hill (854 m.) is the highest point in the area. Some of the other important hills are Hudua (826 m.). Banadag (670 m.), Bambai (784 m.), and Kanari (711 m.). These hills form a ring within which the Hazaribag town is located. This plateau surface descends to Kodarma - Giridih plateau to its north and eastern side with steep scarps. In the southern side it overlooks the Damodar trough. The general slope of the plateau surface is less than 2^0. But it increases near the hills, valleys and on the margins of the plateaus. The relative relief (Fig.17) Varies from less than 25 meters to more than 100 meters near hills and scarp.

This surface is the source region of the several streams. The Barakar, the Konar, the Siwani, the Bokaro, the Haharo and the Mohana (outside the area under study) are some of the important rivers. The pattern and the direction of the rivers are greatly controlled by joints in the rocks and slopes of the area.

One important aspect of the area is the gully erosion. Gullies form the miniature landforms. Severe gully erosion may be observed along the banks of the rivers and on the foot of the hills, etc. Gullies are the universal feature of the entire Basin. But they differ in their extent and magnitude from the one part to another part of the Basin.

4.3. The Inter-Plateau Ridge :

The general level of the Chandwa - Tori - Balumath ridge is 600 metres. Numerous hills e.g. Latdag (1011 m), Chetag (952 m) and Kasiatu (807 m.) rise above the general surface of the area. This ridge is scarped both towards the west by the Auranga valley and towards the east by the Damodar valley. It connects the Ranchi and the Hazaribag plateaus. This ridge is made up entirely of the granites and gneisses and forms an integral part of the Rachi and the upper Hazaribag plateaus. Several streams originate from this ridge. The Dharmauti, the Ramghat, the Dagdagia, the Khobad, the Domohani and the Ranikhet are some of them. They are actively engaged in changing the line of divide between the tributaries of the Damodar and the Auranga. It is peculiar to note that the streams which originate from the eastern slope of Latdag hill (1011 m.) turn west to become the tributaries of the Auranga river through Tori gap. Here the line of divide between the Damodar and the Auranga does not run through the line of the highest elevation. The scarp is relatively gentle. The average slope is less than 6^0. The relative relief is less than 100 metres. But it rises upto 400 metrees near the Latdag hill area. This section of the scarp zone is severely gullied. This can be easily observed along the banks of the streams.

4.4. The Kodarma -Giridih Plateau :

The Kodarma - Giridih plateau is delimited in the east by 300 - metres contour. It is the third planation surface. After the early and late Tertiary uplift which gave rise to the Netarhat plateau and the central Chotanagpur plateau respectively, "there was stillstand and recession of scarp. Then there was uplifts particularly of the whole of Chotanagpur plateau excepting the eastern peneplain surface in Manbhum and country east of Rajmahal Hills, by about 700' - 1000' (214 - 305 m.) in Pleistocene to Recent time" (Ahmad, 1958 : 18). The western part is an open rolling plateau but a few isolated hills interrupt the skyline. The Valley of the Usari has the fine open aspect with crumbing peaks scattered over the area.

Plate - 4 : Khandoli Hill near Giridih

The Khandoli, the peculiar sadddle - shaped hill made up of granitoid gneiss, dominates the landscape near Giridih, the wellknown mica - splitting centre. Its peak is like a volcanic cone and the great saddle attached to it consists of massive boulders (Bose, 1948 : 14).

On Bagodar highland area isolated hills diversity the otherwise flat surface especially in the watershed area. This region has maximum development of short steep ridges and isolated hills.

4.5. The Outer Erosion Fringe :

The outer Erosion Fringe lies in the eastern part of the Basin. It consists of granites and gneisses of Archaeans and sandstone and shales of Gondwana. The surface is probably the result of the last Himalayan upheaval. The area is described as the first step of gradual descent from the table land of Chotanagpur to the delta of Bengal. The country is more open and presents the appearance of a series of rolling downs dotted here and there with isolated residual hills. Most of the hills are conical in shape. The Parasnath (1366 m.) is the highest point in the Basin. "The common types of rocks which give rise to outstanding hill are the meta morphic rocks like epidiorite, amphibolites, metadolerites and metanorites. Some of these hills may thus be regarded as relict types of ridges which have suffered less erosion than the surrounding area. The veins of quartz (often brecciated) which are so common in their metamorphic terrain of the area, also form low ridges may, however, be regarded as tectonic in origin as they usually indicate fault zone" (Sharma, D.G. Dhanbad, Secretariat Press, Bihar, Patna, p. 12).

The relief is less than 25 metres. But the highest relief more than 700 metres in a grid of 1 mile ^{2} | (2.59 Km^{2}) is observed in the Parasnath area.

4.6. The Damodar Valley :

The landscape of the Damodar valley is hetrogenous terrain. It is an uplifted and re - excavated feature in the softer Gondwanas. The influences of structural patterns and rock resistances of erosion are clearly seen in the topological evolution of the area. The western part of the valley is more hilly than the plantation surface. The hard and resistant Gondwana sandstones have given rise to bold topography. Morcha Pahar in the Karnpura Coalfield and Lugu

Pahar in the Bokaro Coal field are entirely made up of hard and resistant sandstones. The shales and softer sandstones form the lowlands.

The landscape from Danea to Bermo is dominated by accidented relief. The Bokaro river forms a magnificent gorge between the Lugu Pahar in the south and Jhumara Pahar on the north.

The landscape of Kathara - Jhirki area is dominated by hills and ridges formed entirely of massive sandstones. Kathara Pahar and the Bandh Pahar in the northern periphery of Bokaro Coal field are notable example. These hills and ridges are, however, not continued along the southern periphery due to interruption by 'oblique faults' and are often absent, especially in Jarangdih, Jaridih, Chalkari and Phusaro areas. Some of the hills are also made up of burnt shale. A hill made entirely of indurated shales can be seen on the north of the Jarangdih railway station. In Jharia and its surroundings the surface looks black due to abundance in coal mining.

The general slope (Fig. 13) of the valley ranges from less than 1^0 in the eastern part to more than 6^0 in the western part. The highest average slope (11^0 43') may be observed near Lugu Pahar. The Morcha Pahar area has an average slope of 9^0 58'. The southern flank of the valley is steeper than the northern flank.

The local relief varies from less than 25 metres in a grid of 1 $mile^2$ (2.59 $Km.^2$) to more than 400 metres.

Faults have greater control over the development and pattern of drainage (fig. 13C). The Damodar itself is a fault guided stream.

The above discussions conclude that physiographic landscape of the area is the result of the latest phase of the tectonic uplift of the Basin and subsequent subaerial denudation. The evolution and the present characteristics of the landforms are largely controlled by the structural pattern and resistances of rocks to their erosion. Bold topography (e.g. hills, ridges and scarps) are generally associated with hard and resistant - rocks. The softer rocks form the lowlands. The planation surface are the rejuvenated peneplains, etc.

5

Morphometric Analysis of the Linear Aspects

5.1. Morphometric Analysis :

Morphometry is defined as "the measurement and mathematical analysis of the earth's surface and of the shape and dimensions of its landforms" (Clarke, 1958 : 235). During the last seventy years, subjective approach to the description of drainage basin has been supplanted by objective and useful techniques of study. These perhaps form the bases of drainage morphometry.

Horton (1945) initiated morphometric techniques for the systematic and quantitative analysis of erosional landforms in the 'water shed' area. Horton's work was developed in detail by Strahler (1952, 1954, 1956, 1964), Schumm (1956), Melton (1957), Chorley (1958), Morisawa (1959) and several others.

In India, the application of morphometric techniques in classification of landforms was initiated by R.L. Singh (1967). Since then numerous young research workers are encouraging morphometric techniques in geomorphic study of drainage basins.

The main objective of drainage basin morphometry is to gather accurate data of measurable features of stream networks and drainage basins. These data have been analysed for the establishment of certain basic laws of stream behaviour and landscape evolution, etc.

Morphometric techniques have been applied for the quantitative expression of the linear, areal, and relief aspects of the Damodar Basin in Jharkhand. Attempts have also been made here to present a factual and accurate picture of the morphological aspects of the Basin.

5.2. Linear Aspects of the Channel System :

The linear aspects deal with one dimension of the stream segments. Hence, order - number, order - length, meandering and Sinuosity Indexes have been included in this chapter.

5.2.1. Stream Order :

"Stream order is a measure of the position of a stream in the hierarchy of tributaries" (Leopold, et al. 1964 : 134). This offers "a quantitative basis for comparision of the degree of development in the drainage nets of comparable size" (Horton, 1945 : 283).

The network of the channels of the Damodar Basin in Jharkhand has been traced from the quarter inch topographical sheets and divided into hierarchical orders of different magnitudes (Fig. 4.), following Strahler's method, which is a slight modification of Horton's law of stream order. According to Strahler's modification the difficulty of re - numbering of streams is overcome, the number of streams increased and the lenght of the main channel shortened. The orders have been designated and orderwise number of streams have been counted. Bifurcation ratio (Rb) has been calculated (Table. 5 - 1). The data have been plotted (Figs. 5A, 5B and 5C), taking constant - ratio scale for numbers and arithmetic scale for orders.

TABLE 5 - 1

Basin Morphometry : Order - Number

A. THE DAMODAR

Stream order, U	Number of Stream segments, Nu	Bifurcation* Ration, Rb
1st.	1905	3.14
2nd.	460	
3rd.	119	3.86
4th.	29	4.10
5th.	8	3.62
6th.	3	2.66
7th.	1	3

Av. Rb. 3.39

B. THE BARAKAR

Stream order, U	Number of Stream segments, Nu	Bifurcation* Ration, Rb
1st.	801	
2nd.	189	4.23
3rd.	50	3.78
4th.	9	5.55
5th.	2	4.50
6th.	1	2.00

Av. Rb. 4.01

C. THE KONAR

Stream order, U	Number of Stream segments, Nu	Bifurcation* Ration, Rb
1st.	222	
2nd.	44	5.04
3rd.	12	3.66
4th.	5	2.40
5th.	2	2.50
6th.	1	2.00

Av. Rb. 3.12

Source : Quarter - Inch Topographical Sheets.

$$* \text{Rb} = \frac{\text{Nu}}{\text{Nu} + 1}$$

Where Rb = ———— Bifurcation Ratio

Nu = number of stream segments of order U

Figures and Table reveal the following facts :

(1) The Damodar in Jharkhand is a 7th order Basin. There are 3 sixth, 8 fifth, 29 fourth, 119 third, 460 second and 1905 first order - basins.

(2) There is a progressive decrease in the number of streams with the increase in the value of stream orders. The plotted data produce a negative exponential form.

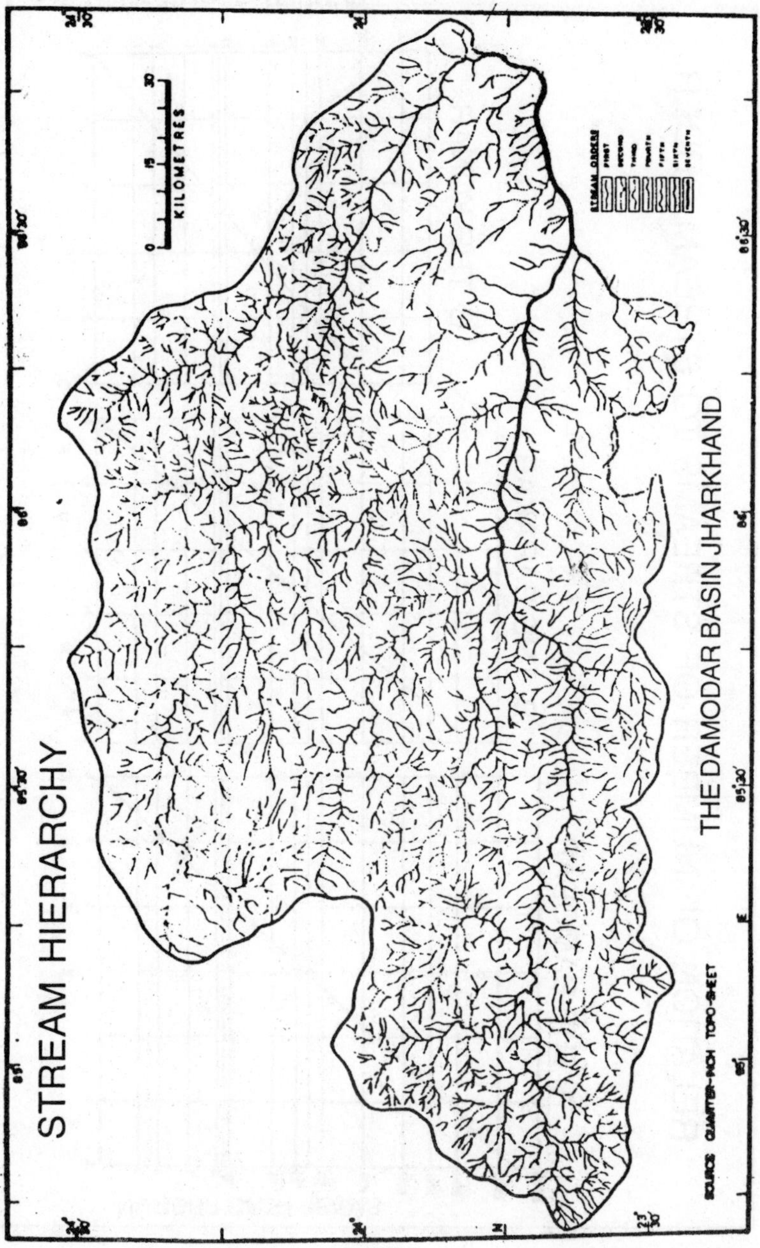
STREAM HIERARCHY
THE DAMODAR BASIN JHARKHAND
KILOMETRES
STREAM ORDERS

Fig. 4

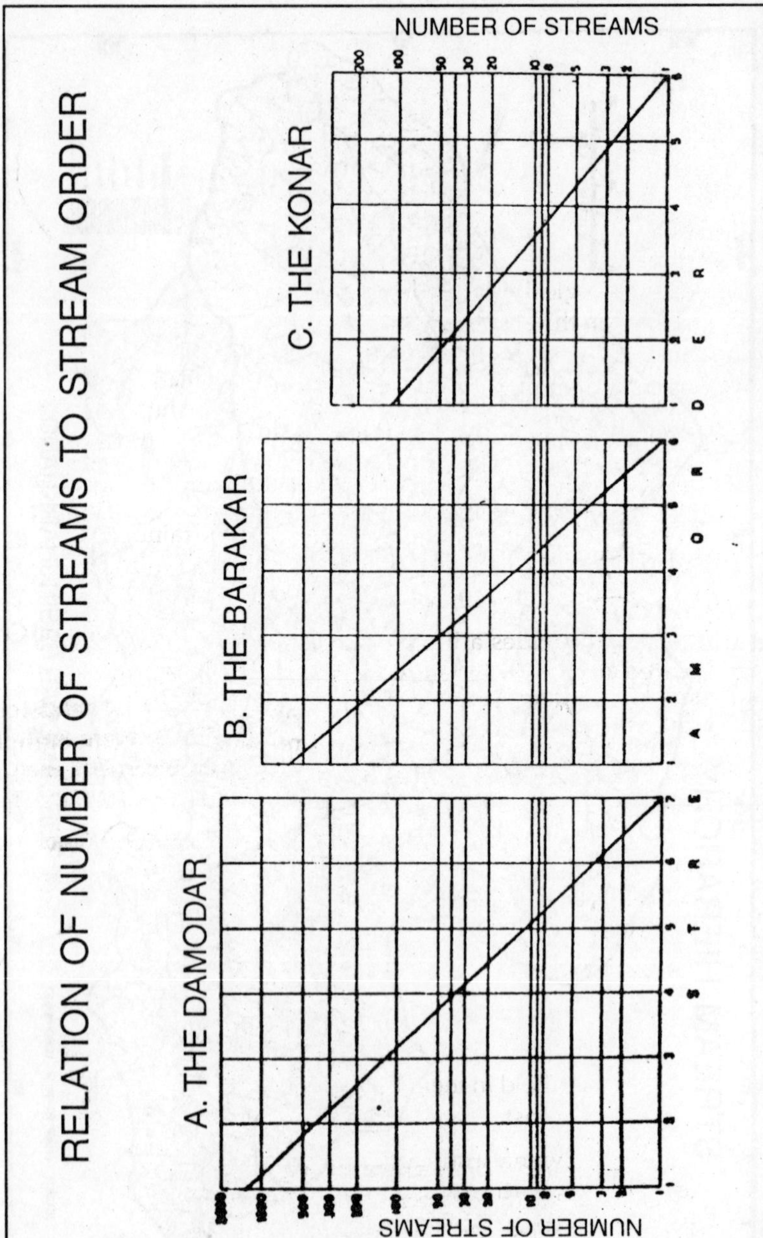
RELATION OF NUMBER OF STREAMS TO STREAM ORDER
A. THE DAMODAR
B. THE BARAKAR
C. THE KONAR
NUMBER OF STREAMS
NUMBER OF STREAMS
STREAM ORDER

Fig. 5

(3) Bifurcation ratio, (Rb) varies from 2 to 5.5. It is minimum in case of the Barakar and the Konar basins 5th and 6th orders. Maximum bifurcation ratio, (Rb) has been observed between 3rd and 4th order basins of the Konar.

(4) The probable explanation for the variation in bifurcation ratio, (Rb) is due to variations in the nature and character of slopes, complex geological structure and epeirogenetic uplift of the Basin. Numerous finger - tip streams have been developing in the Basin due to high rate of erosion, giving higher order value to older stream segments. This also may be responsible for the variation in the bifurcation ratio.

5.2.2.Stream Lenghts :

Orderwise lengths of stream segments of the Damodar Basin in Jharkhand have been measured with the help of rotameter. Mean langths (Lu), cumulative mean lengths and length - ratio (RL) have been calculated (Table 5 - II). Data have been plotted (Figs. 6A, 6B, 6C,) taking constant - ratio scale for mean and cumulative lenghts and arithmetic scale for orders of stream segments.

TABLE 5 - II

Basin Morphometry : Order - Length

A. The Damodar Basin in Jharkhand

Order U	Lu in Kms.	Mean lengths Lu*	Cumulative mean lengths C Lu	Length Ratio Rl**
1	2	3	4	5
1st.	5149.76	2.7	2.7	
2nd.	1564.84	3.4	6.1	1.25
3rd.	812.69	6.83	12.93	2
4th.	534.28	18.42	31.35	2.69
5th.	244.61	30.57	61.92	1.65
6th.	194.72	64.90	126.82	2.12
7th.	109.43	109.43	236.25	1.68
				Av. R1 = 1.89

B. The Barakar Basin

1st.	8182.21	2.72	2.72	1.2
2nd.	659.81	3.49	6.21	1.28
3rd.	302.54	6.05	12.26	1.73
4th.	241.39	26.82	39.08	3.43
5th.	104.6	52.3	91.38	1.95
6th.	70.8	70.8	162.18	1.35

Av. R1 = 2.15

C. The Konar Basin

Order U	Lu in Kms.	Mean lengths Lu*	Cumulative mean lengths C Lu	Length Ratio Rl**
1	2	3	4	5
1st.	645.35	2.906	2.906	1.22
2nd.	156.1	3.547	6.453	2.04
3rd.	86.9	7.241	13.694	
4th.	59.5	11.9	25.594	1.64
5th.	72.41	36.205	61.799	3.04
6th.	8.04	8.04	69.839	0.22

Av. R1 = 1.632

Source : Quarter Inch topographical sheets.

$$*\bar{L}u = \frac{\bar{L}u}{Nu}$$

Where

$\bar{L}u$ = Mean length of channel segments of order U.

Lu = Total lengths of stream segment of order U.

Nu = Total number of stream segment of order U.

$$**RL = \frac{\bar{L}u}{\bar{L}u - 1}$$

Where

RL = Length ratio.

The Table 5 - II and the Figs. 6A, 6B, 6C, reveal the following facts:

(1) The mean lengths (Lu) of the first order streams is the shortest. It increases with the increase in the value of the order (U) of the stream segments.

(2) But in case of the Konar, the mean length 8.04 Kms. of the 6th order channel is far less than the mean lengths 36.2 Kms. of the 5th order channel segments. (Table 5 - IIC and Fig. 6C).

The probable explanation is that the Konar is a tributary of the fault guided Damodar. It joins the Damodar immediately after coming out of scarpland topography. Here, the Damodar also turns to receive the Konar river, which is its important tributary. The topography or slope is the controlling factor for the same.

(3) The plotted data for the Damodar and the Barakar (Figs. 6A, 6B) produce a positive exponential form. But all points do not fall perfectly on the straight line.

(4) The point of the 6th order of the Konar (Fig. 6C) is far away from the straight line.

(5) The average length ratios (RL) for the Damodar, the Barakar and the Konar are 1.89, 2.15 and 1.63 respectively. But they vary from 1.28 to 3.43 in case of the Barakar and 0.22 to 3.04 in case of the Konar.

The variation in length - ratio (RL) may be explained as due to variations in geological structures and topography in the area. This probably discloses the clue that the area has been rejuvenated. The development of numerous new stream segments perhaps is responsible for the variation of lenght ratio (RL).

5.2.3. Meandering and Sinuosity Indexes :

5.2.3.1. Meandering :

Meandering may be defined as the winding nature of the course of a river. As a rule "no straight river occurs in a state of nature" (Davis, 1913 : 4) all streams are characterised by meandering and sinuous courses. Thus, meandering is "a natural state of affairs, obtained by gradual stages as the channel departs from a straight line" (Tanner, 1968 : 959). Straight channels are so rare "as to be almost non - existent" (Leopold and Wolman, 1957).

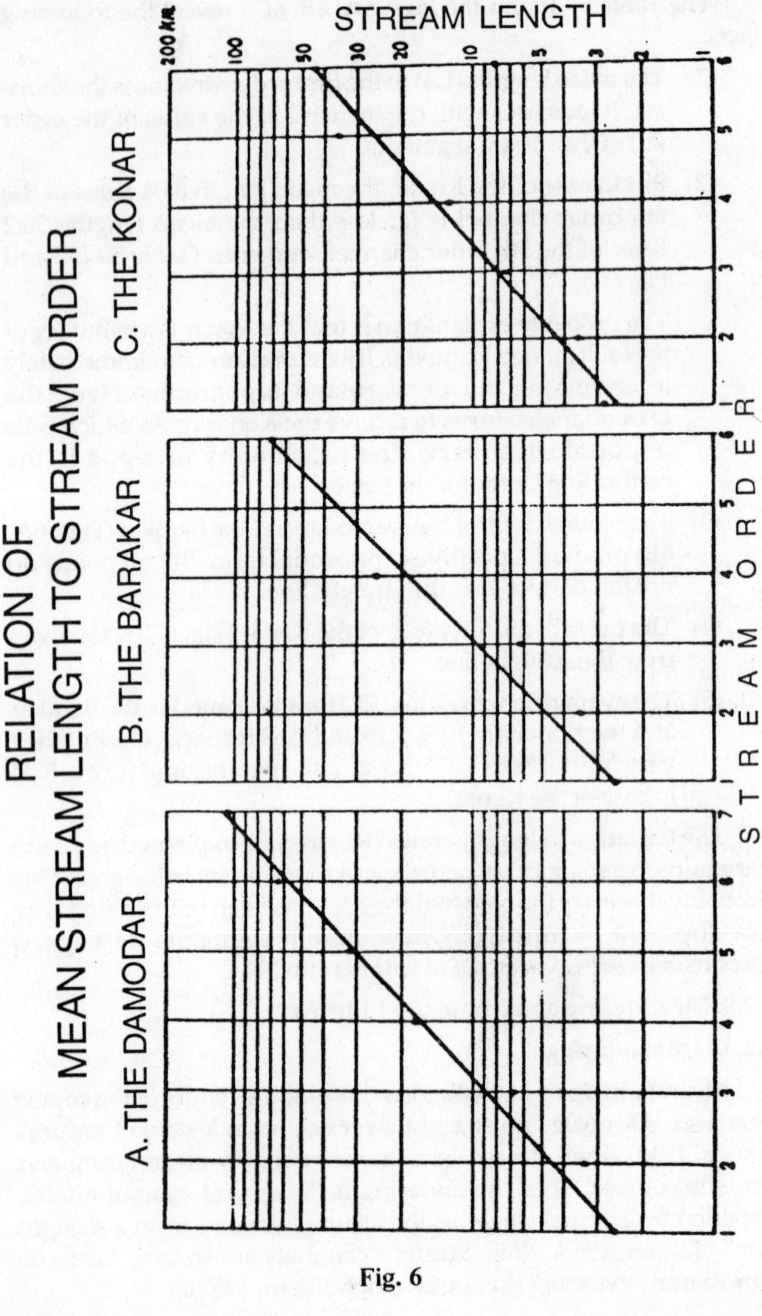
STREAM LENGTH
200 KM
100
50
30
20
10
5
3
2
1
RELATION OF
MEAN STREAM LENGTH TO STREAM ORDER
A. THE DAMODAR
B. THE BARAKAR
C. THE KONAR
STREAM ORDER
1
2
3
4
5
6
7

Fig. 6

Plate - 5 : Meandering Channel of Konar

Meandering streams have traditionally been ascribed to fluvial plains and delta plains (Thornbury, 1954). From statements of this kind has arisen the notion that meandering is limited to large rivers which are mature, in the Davisian sense or supemature or "old" (Tanner, 1968 : 958). It seems to be incorrect, because in the Damodar Basin in Jharkhand streams of all size meander at all altitudes.

In the meandering courses of the streams both the processes of erosion and deposition may be observed. Bank erosion may conspicuously be seen on the concave side where the velocity is stronger and sediments are accumulated at the convex bank, where water is less deep and flows slower.

In the Damodar Basin in Jharkhand meandering streams have not developed on wide flood - plains. They flow in comparatively deep valleys with steep banks. They are actually entrenched, which flow through "a gorge or chasm cut in bed rock" (Tanner, 1968:962). We have an excellent example at Rajrappa, about 16 Kms. north of Gola in the Damodar river. This has been cut down into bed rock because of the epeirogenetic uplift of the area.

Incision of streams is probably continuing in the different parts of the Basin. This may well be observed in the Usari, the Konar, and the Dhainti sub - catchments of the Basin. It suggests that the area is still rising probably in the form of tilts, which supports the concept of E.Ahmad (1973 :60) that the area is rising.

Plate - 6 : Confluence of Konar of Siwane

Alternatively, if we consider that meandering is related with faults and joints in the rocks of the area, we face difficulty in proving the same. The faults are generally straight. "The strike of the jointed gneiss probably does not turn as sharply and frequently as the meanders " (Ahmad, 1969 :IX). Hence, faults and joints cannot be the guiding factors for the initiation and development of meanders in the Basin.

The above discussion concludes that the genesis of meanders may be explained by assuming the uplift of the Basin epeirogenetically in the recent past and the area is still rising in the form of gentle tilts but the rate at which the area is rising is still to be proved.

5.2.3.2. Sinuosity Indexes :

Meandering may be "applied to sinuous channels that exhibit a certain degree and regularity of sinuosity" (Brice, 1962 : 8). Leopold and Wolman (1957) have defined sinuosity of the streams as the ratio of thalweg length to valley length and also (Leopold &

Longbein, 1960) as the ratio of arc distance to wave length in a single meander. Friedkin (1945) defines sinuosity as the thalweg lenght to air - line distance.

Of all the existing sinuosity indexes, the most common is that which is derived by dividing the length of a reach as measured along the channel by the length of a reach as measured along the valley. According to this index a stream having a sinuosity of 1.O. is straight, just above unity to 1.3 is sinuous and more than 1.3 is meandering (Prasad, 1979 : 193).

These indices give us the idea of the total sinuosity of the streams. Having felt some inherent flaws in these sinuosity indices, Mueller (1968) has derived a new Sinuosity Index. He has given due importance to the topography and the channels in making sinuous bends of the streams. He is of the opinion that topography plays an important role in the development of streams and during the early stage of a cycle of erosion sinuosity of the streams is a direct function of the sinuosity of the valley and additional sinuosity is imminent as the current eats into the outside of each bend or meander and build up the inside, so the course of the river becomes more tortuous. Hence almost all streams have both Hydraulic and Topographic Sinuosity, for it must be remembered that "streams became prisoners of the valleys which they cut and occupy" (Lieghy, 1936). Therefore, it is essential to analyse the influence of topography and hydrology on the development of sinuosity of the streams.*

*** Index of Sinuosity :**

$$\text{HSI (Hydraulic Sinuosity Index)} = \frac{\text{CI - VI}}{\text{CI - 1}} \times 100$$

$$\text{TSI (Topographic Sinuosity Index)} = \frac{\text{VI - 1}}{\text{CI - 1}} \times 100$$

$$\text{CI (Channel Index)} = \frac{\text{CL (Channel Length)}}{\text{Al (Air Length of Straight Length)}}$$

$$\text{VI (Valley Index)} = \frac{\text{VL (Valley Length)}}{\text{AL (Air Length)}}$$

$$\text{SSI (Standard Sinuosity Index)} = \frac{\text{CI}}{\text{VI}}$$

N.B. :

1. CI is an index of total sinuousity, both hydraulic and topographic.

2. VI is an index of total topographic sinuousity.

3. Valley Length (VL), along a stream, is the lenght of a line which is everywhere midway between the base of the valley-walls. It will equal Channel Length (CL) wherever the valley-walls decend directly to the water's edge, and will be less than CL wherever a flood plain has developed.

4. Air Length (AL) is the shortest distance between two points of a meandering bend of a river (for big rivers) and the shortest distance beween the source and mouth of the streams in case of the small reivers.

With these definitions the Topographic Sinuosity Index (T.S.I.) and Hydraulic Sinuosity Index (H.S.I.) of some of the streams of the Basin (as a sample study) have been measured from one - inch topographic Sheets.

From table 5 - III the following points are to be noted :

(1) Hydraulic Sinuosity Index (H.S.I.) varies from 0 0 % to 47.06%. It is minimum at bend no. I of the Damodar and the Bhera rivers and maximum at bend No. 12 of the Damodar river.

(2) Topographic Sinuosity Index (T.S.I.) varies from 52.94 % to 100 %. It is minimum at bend No. 12 of the Damodar and maximum at bend No. 1 of the Damodar and the Bhera rivers.

(3) Standard Sinuosity Index (S.S.I.) varies from 1.00 to 1.08 in case of the Damodar, 1.03 to 1.08 in case of the Konar, and the Jamunia and 1.03 to 1.07 in case of the Bokaro.

(4) The higher percentage of T.S.I. is due to the fact that the topography has a major control over the development of sinuosity in the Damodar Basin in Jharkhand.

(5) The lesser percentage of H.S.I. probably gives a clue that the Basin is passing through youth to maturity, bacause in late

Table 5-III
Sinuosity Index (Mueller's Model)

Streams	V.L. (km.)	C.L. (km.)	A.L. (km.)	C.I.	V.I.	H.S.I. %	T.S.I. %	S.S.I.	Gradient M./km.	Average Width in metres
1	2	3	4	5	6	7	8	9	10	11
The Damodar										
Bend I (A-B)	10	10	8.8	1.14	1.14	00.0	100.0	1	55.81	18.6
Bend II (B-C)	5.6	6	4.4	1.36	1.27	25.0	75.0	1.07	4.22	85.88
Bend III (C-D)	40.8	43.6	28.8	1.51	1.42	17.6	82.4	1.06	1.87	165.85
Bend IV (D-E)	26.8	28	16	1.75	1.67	10.67	89.33	1.05	1.52	182.5
Bend V (E-F)	25.6	27.2	18.4	1.48	1.39	18.75	81.25	1.6	0.56	179.44
Bend VI (F-G)	11.2	12	8	1.5	1.4	20.0	80.0	1.07	2.03	165.65
Bend VII (G-H)	33.8	36	27.2	1.32	1.24	25.0	75.0	1.06	1.38	154.13
Bend VIII (H-I)	30.4	32	25.6	1.25	1.17	32.0	68.0	1.07	1.86	386.49
Bend IX (I-J)	32.8	35.2	27.2	1.29	1.2	31.04	68.96	1.08	1.32	410.6
Bend X (J-K)	16.8	18	15.2	1.18	1.10	44.44	55.56	1.07	1.32	344.6
Bend XI (K-L)	21.6	22.8	17.6	1.29	1.23	20.68	79.32	1.05	0.51	445.54
Bend XII (L-M)	36.8	39.2	33.6	1.17	1.09	47.06	52.94	1.07	0.77	748.84
The Konar										
Bend I (N-O)	13.6	14	12	1.16	1.13	18.75	81.25	1.03	4.66	37.57
Bend II (O-P)	11.8	12.4	9.2	1.34	1.28	17.65	82.35	1.04	1.13	69.78
Bend III (P-Q)	19.2	20.8	14.4	1.44	1.33	25.0	75	1.08	2.63	138.49

Table contd..........

1	2	3	4	5	6	7	8	9	10	11
Bend IV (Q-R)	9.8	10.4	6.4	1.62	1.53	14.51	85.49	1.06	2.05	85.88
Bend V (R-S)	31.4	33.2	17.6	1.86	1.78	9.3	90.7	1.04	3.49	135.27
Bend VI (S-T)	22.4	23.6	18.4	1.28	1.22	21.42	79.58	1.05	6.72	152.98
Bend VII (T-V)	6.4	6.8	4.8	1.41	1.33	19.51	80.49	1.06	1.75	203.98
The Jamunia										
Bend I (V-W)	14.0	14.4	10.8	1.33	1.29	12.12	87.88	1.03	10.48	17.5
Bend II (W-X)	8.0	8.4	6.4	1.31	.125	19.35	80.65	1.04	4.94	26.6
Bend III (X-Y)	13.2	14.4	8.8	1.63	.15	20.63	79.37	1.08	2.02	30.4
Bend IV (Y-Z)	15.2	16.0	10.4	1.53	1.46	12.28	87.72	1.04	3.62	33.5
Bend V (Z-Z[1])	34.8	27.2	22.4	1.66	1.55	16.66	83.34	1.07	1.64	50.3
Bend (Z[1]-Z[2])	17.6	18.4	13.6	1.35	1.29	17.14	82.86	1.04	5.52	67.1
The Bokaro										
Bend I (a-b)	17.2	17.76	14.8	1.2	1.16	20.0	80.0	1.03	10.39	16.0
Bend II (b-c)	12.2	12.8	8.4	1.52	1.45	13.46	86.54	1.07	4.76	44.28
Bend III (c-d)	15.2	16.0	9.6	1.66	1.65	1.51	98.49	1.06	4.00	61.85
Bend IV (d-e)	13.6	14.4	11.2	1.28	1.21	25.0	75.0	1.05	2.88	106.03
Bend V (e-f)	12.8	13.6	11.2	1.21	1.18	14.29	85.71	1.03	2.55	140.91
Bend VI (f-g)	14.2	15.2	11.2	1.35	1.27	22.85	77.15	1.06	2.2	118.09

N.B. :

V.L. = Valley Lenght; C.L. = Channel Lenght; A.L. = Air Length; C.I. = Channel Index = CL/AL; V.I. = Valley Index = VL/AL; H.S.I. = Hydraulic Sinuosity Index; T.S.I. = Topographic Sinuosity Index; S.S.I. Standard Sinuosity Index;

A = Source of the Damodar; B = Damodar bridge; C= Confluence point of the Damodar and the Dharmauti; D = Confluence point of the Damodar and the Saphi; E = Confluence point of the Damodar and the Haharo; F = Confluence point of the Damodar and the Marmarha; G = Confluence point of the Damodar and the Konar; J = Confluence point of the Damodar and the Jamunia; K = Confluence point of the Damodar and the Katri; L = Confluence point of the Damodar and the Gobai; M = Confluence point of the Damodar and the Barakar.

N = Source of the Konar; O = Confluence point of the Konar and the Gonda; P = Near Hathigaon; Q = Near Chapodhurd; R = Confluence point of the Konar and the Siwani; S = Near Mahto tolo; T = Confluence point of the Konar and the Bokaro; U = Confluence point of the Konar and the Damodar.

V = Source of the Jamunia; W = Near Bishnugarh; X = Pandedih; Y = Near Balak; Z = Confluence point of the Jamunia and the Kulgo; Z1 = Near Railway Bridge; Z^2 = Confluence point of the Jamunia and the Damodar.

a = Source of the Bokaro; b = Near Waterfall; c = Confluence point of the Bokaro and the Dudhi; d = Confluence point of the Bokaro and the Chotha; e = Confluence point of the Bokaro and the Chutua; F = Near Dewan toli; g = Confluence point of the Bokaro and the Konar.

maturity or in old stage the control of water should be greater than topography and in that case the percentage of H.S.I. should be higher than the percentage of T.S.I.

(6) This may also be probable that Sinuosity is being adjusted to the old river course after the epeirogenetic uplifts of the area.

(7) In the upper section of the Basin higher percentage of T.S.I. sometimes 100 % indicate total control of topography on the development of Sinuosity, may be the feature of youth stage. In the lower section the relative decrease in the percentage of T.S.I. is the indicative of the increase influence of hydrology on the development of Sinuosity, may be the feature of maturity and hence we may conclude that youth and maturity occur hand in hand in the Basin.

6

Morphometric Analysis of the Areal Aspects

6.1. Areal Elements :

The area of a drainage basin of order (U) is defined as "the total area of surface contributing to all first order channels plus all included inter - basin areas" (Strahler, 1971 : 487). For example, the area of seventh order (Au 7), Basin of the Damodar, would cumulate the areas of all first, second, third, fourth, fifth and sixth order basins plus all additional surface elements, known as inter - basin areas, contributing directly to the channel of higher order.

6.2. Frequency Distribution :

Frequency distribution of areas of drainage basins has been studied by V.C. Miller (1953 : 14) and Schumm (1956 : 607). Horton (1945 : 294) inferred that mean drainage - basin areas of progressively higher orders should increase in a geometric sequence, as do stream lengths. Schumm expressed this relationship in a law of stream areas; the mean basin areas of streams of each order tend closely to approximate to a direct geometric sequence in which first term is mean area of the first order basin (Au 1). Morisawa, (1959 : 51) confirmed this relationship. Leopold and Miller (1956, 19 - 20) found the law to apply to the basins of ephemeral streams in central New Mexico.

For areal analysis of the Damodar Basin in Jharkhand, the perimeters of all first, second and higher orders of basins have been drawn on topographical map. The orderwise areas have been measured. Mean area (Au) and area - ratio (Ra) have been calculated (Table 6 - I). Orders and areas have been plotted (Fig. 7) taking constant - ratio scale for areas and arithmetic scale for orders.

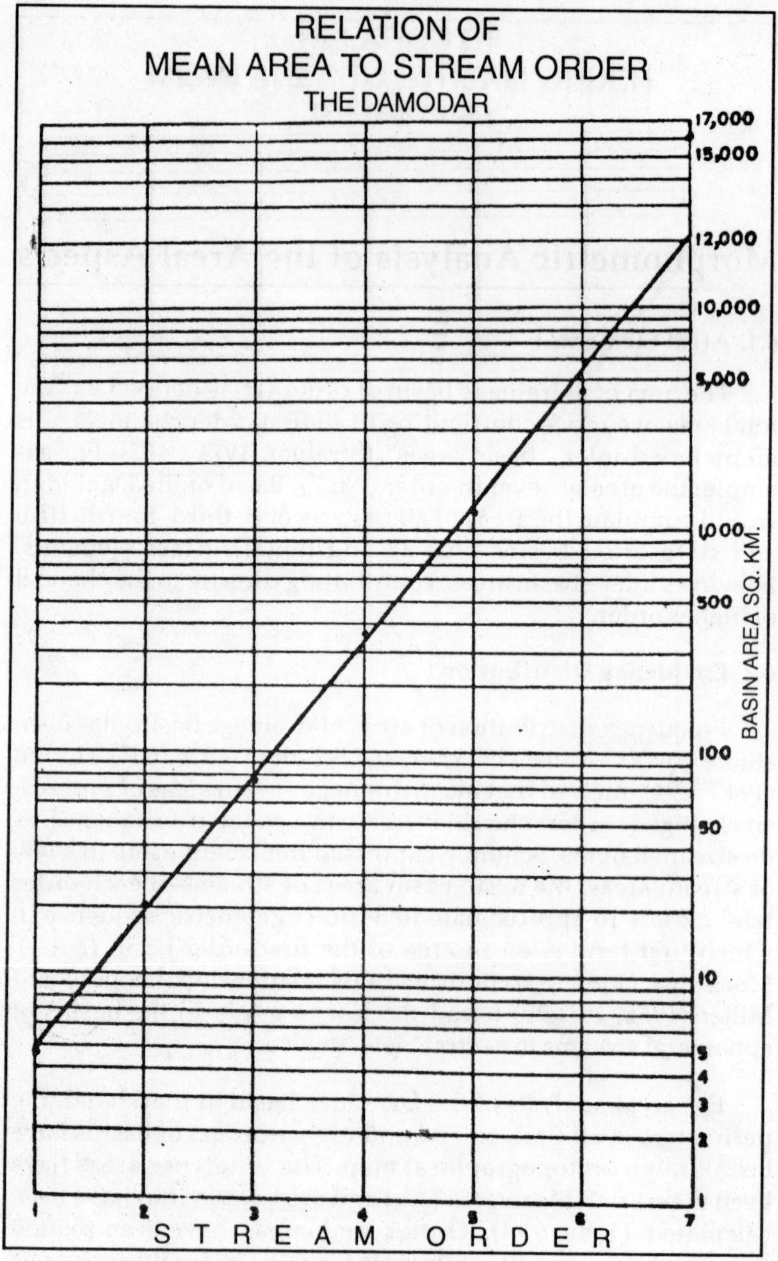
RELATION OF
MEAN AREA TO STREAM ORDER
THE DAMODAR
17,000
15,000
12,000
10,000
5,000
1000
500
100
50
10
5
4
3
2
1
BASIN AREA SQ. KM.
1
2
3
4
5
6
7
STREAM ORDER

Fig. 7

TABLE 6 - I

Basin Morphometry : Order - Area

THE DAMODAR BASIN IN JHARKHAND

Order U	Total Area km^2	% of total areas	Mean area	Area Ratio (Ra)
1st.	9430.99	55.71	4.95	
2nd.	9611.10	56.78	20.89	4.22
3rd.	9105.81	53.79	76.52	3.66
4th.	10986.88	64.91	378.86	4.59
5th.	10935.92	64.61	1366.99	3.81
6th.	13898.98	82.12	4632.00	3.39
7th.	16249.66	100%	16249.66	3.51

Average Ra = 3.83

Table 6 - I and Fig. 7 reveal the following facts :

(1) Mean area (Au) increases with increase in the value of the order of the Basin.

(2) Mean area of the basins of each order, plotted against order (Fig. 7) produces a straight line regression of positive exponential form.

(3) The plotted data do not show perfect alignment with the straight line. Points of the second, fourth and seventh orders are not on the straight line but they fall near it and hence the result may be said good to a great extent.

(4) The average area ratio (Ra) is 3.83. But it varies from 3.39 to 4.59. It is minimum between the fifth and the sixth order basins, and the maximum between the third and the fourth order basins (Table 6 - I).

The probable explanation for the variation in area ratio (Ra) may be noted as follows :

1) The erosion rate is high. As new channel segments come into existence, orders of older channel segments have assumed higher orders.

2) River piracy may also explain this variation in area ratio.

3) This may also be explained as due to interruption in the development of landscape resulting into higher frequency of drainage numbers.

4) This variation in Ra strongly suggests the rejuvenation of the area. The Basin is perhaps, still rising.

6.3. Run - Off :

Run - off consists of "all surface - water flow, both over the land slopes and in streams" (Strahler, 1971 : 415). Run - off in the Basin is derived directly from excessive precipitation which cannot infiltrate into the soil, or it also originates as the outflow of ground water along the line where the water table intersects the earth's surface.

It is a subject of vital concern as a basic natural resources upon which the agricultural and industrial development of the Basin are heavily based. Run - off held in reservoirs behind dams provides water supplies for hydroelectricity generation, irrigation and for several other purposes.

It also creates some problems in the Basin e.g. accelerated erosion in the upper parts, siltation in the reservoirs constructed for hydel power generation, irrigation, etc. and flood in the lower parts of the Basin. It is a potent agent of erosion and engaged in transporting the eroded materials and changing the face of the Basin.

It is controlled mainly by the nature and character of rocks, the slopes, the vegetation and the precipitation. The nature and character of rocks and the relief have been discussed in the previous chapters. Here, it is essential to note that the terrain in the Basin as a whole is "ungraded causing rapid drainage, floods and erosion" (Ahmad, 1965a : 103).

The Basin is covered with rich forest, especially in the scarp zone and hilly tract and with sparse vegetation relatively in the densely populated areas. It has its own role to play in controlling the run - off. The roots of trees make space for the infiltration of water. Vegetation also checks free movement of water. Consequently water moves down the slope very slowly and is retained for a longer duration. But, fire and mismanagement are reducing the density of forest resulting into quick discharge especially in the rainy season and on marked slopes.

Precipitation in the Basin is in the form of rain, at times in downpours. Most of it is derived in the season of general rains (from June to October). This is generally within a period of 58 to 76 days. It means that 289 to 307 days in a year are without rain. Even in these rainy days the intensity of rainfall is not uniform. Thus, this variation in rainfall is mainly responsible for the variation in run - off in the Basin.

It reveals the following facts regarding the distribution of rainfall in the Basin.

(1) The total annual rainfall ranges from 1000 mm. to more than 1500 mm. in the Basin.

(2) The distribution of rainfall appears to be complicated giving rise to a number of pockets of high and low rainfall, perhaps, depending on the peculiarities of the terrain.

(3) Two pockets of very high rainfall appear to be conspicuous. The high rainfall at Parasnath is surrounded by relatively low rainfall.

(4) In the southern part of the Basin, the rainfall increases, probably, due to rise of level.

(5) Jamua (1030.7 mm.), Chas (1128.4 mm.) and Tandwa (1230.9 mm.) are the centres of low rainfall.

The general rainy months of the Basin receive about 80 per cent of the annual rainfall. All days of these months are not rainy days. Only 44 to 58 days are rainy during these monsoon months. Out of these certain days are characterised by heavy down pour. The stormy rainfall, whenever it comes, is followed by increased run - off and flood. This is why, "most of the smaller tributaries of the Damodar had to carry water beyond their channel capacity because of heavy down pours" (S.P. Chaterji, 1969 : 15).

The 1959 flood was particularly disastrous as in that year at least six rain storms visited the Basin, causing peak flows at Durgapur of the order of 23000 Cubic metres second.

TABLE* 6 - II
Average Run - Off In Centimetres (1901 - 1950)

	June	July	August	September
Dhanbad	3.12	22.78	22.12	9.09
Dumari	0.86	17.27	21.03	8.08
Bagodar	2.49	19.81	16.54	7.95
Giridih	3.43	19.30	17.40	7.85
Dhanwar	1.47	19.46	17.04	6.58
Hazaribag	5.08	19.41	21.74	9.53
Tandwa	1.09	21.41	20.64	9.17
Ramgarh	4.52	21.29	19.89	6.73
Gola	7.21	24.49	20.85	6.60
Balumath	2.72	24.16	28.17	9.58

Source : Planning Atlas, Damodar valley p. 14.

It is clear from the table that run - off in the month of June varies from 0.86 cm. to 7.21 cm. The run - off line of 2.5 cm. is roughly coincides the 175 mm isohyetal line.

In July and August higher rainfall causes maximum run - off in the Basin.

With decrease in the amount of rainfall in September, the rivers begin to dry up and during the summer months from March to May most of the streams dwindle away .

The general slope of the Basin (Fig.15) is towards the east. The run - off increases towards the east because of the increase in the area of the catchments.

The total water resource of the Basin is derived from 1250 mm. of rain falling on a catchment of 16249.65 sq.Km. "The annual fall delivers roughly 3.5 million tons. of water on every square mile" (2.59 Km.^{2}) (Report on soil conservation working plan, D.V.C. Hazaribag, 1972, p.20).

Most of the streams are ephemeral in character. The stream discharge fluctuates very much from base flow to peak discharge ranging from 2 - 3 cusecs (less than 1 cumecs) to 20,654 cusecs (585 cubic metre second) Table 6 - III. In the year 1913 peak discharge was

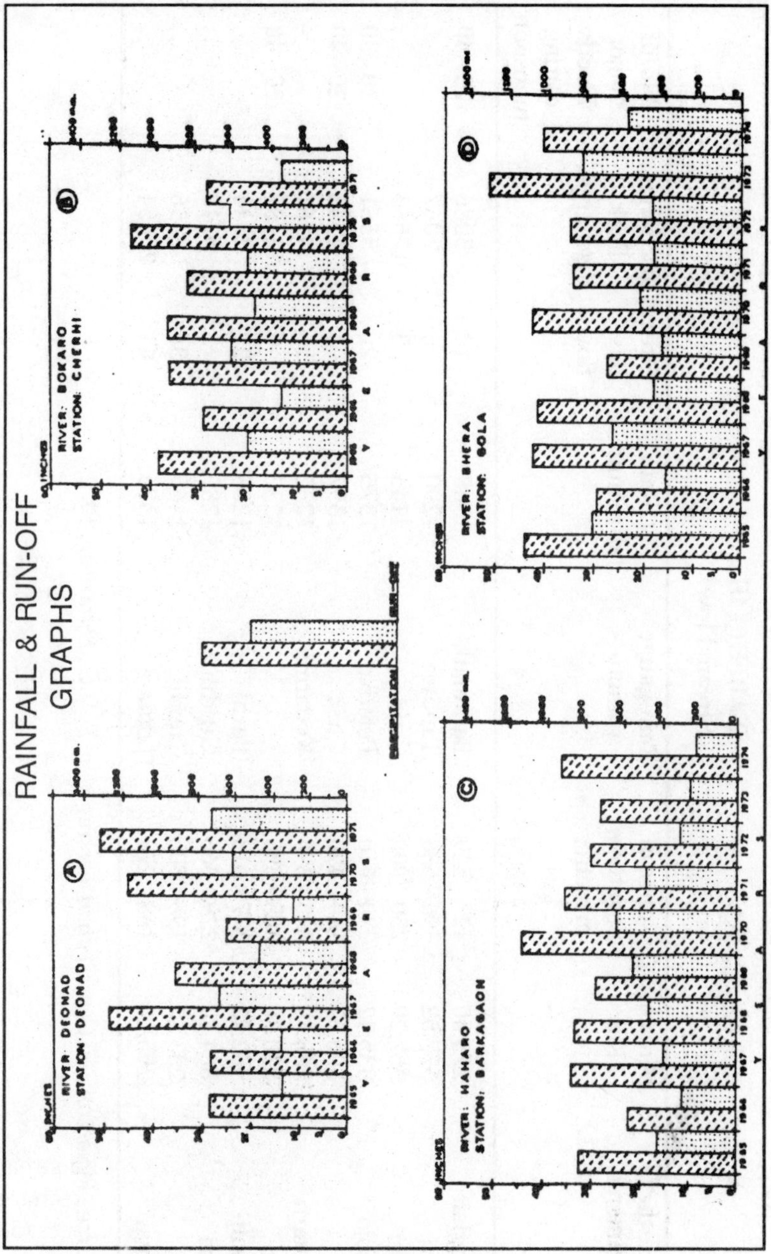
RAINFALL & RUN-OFF GRAPHS
RIVER: DEONAD
STATION: DEONAD
RIVER: BOKARO
STATION: CHERHI
RIVER: HAHARO
STATION: BARKAGAON
RIVER: BHERA
STATION: GOLA
PRECIPITATION

Fig. 8

TABLE 6 - III
Stream Flow*

Name of catchments.	Area in Km.	Range of elevations in metre.	Rain-gauge station.	Av. Rain fall in mm.	Stream discharge		No. of major Floods during monsoon
					Base flow	Peak discharge	
Ramghat	273.50	427 - 549	Balumath	1369	12	5086	20 - 40
Barki	577.52	396 - 686	Tandwa	1290	6	7067	,,
Bhera	357.29	230 - 900	Gola	1143	3	12563	,,
Khanjo	335.12	230 - 610	Peterbar	1375	20	5561	20 - 30
Ijri	503.37	140 - 420	Chas	1397	3	4142	20 - 40
Harnaro	383.19	335 - 396	Kodarma	1232	56	11161	-
Keso	220.20	305 - 396	,,	1232	9	12487	20 - 40
Bansoti	582.22	377 - 613	Barhi	1286	22	2344	,,
Usari	864.20	229 - 366	Giridih	1289	3	5749	,,
Beri	233.12	198 - 305	Tundi	1465	36	8356	,,
Rajoya	295.20	168 - 274	Jamtara	1376	41	20654	,,

*Source : *Report on Soil Conservation working plan, D.V.C., Hazaribag*, 1972.

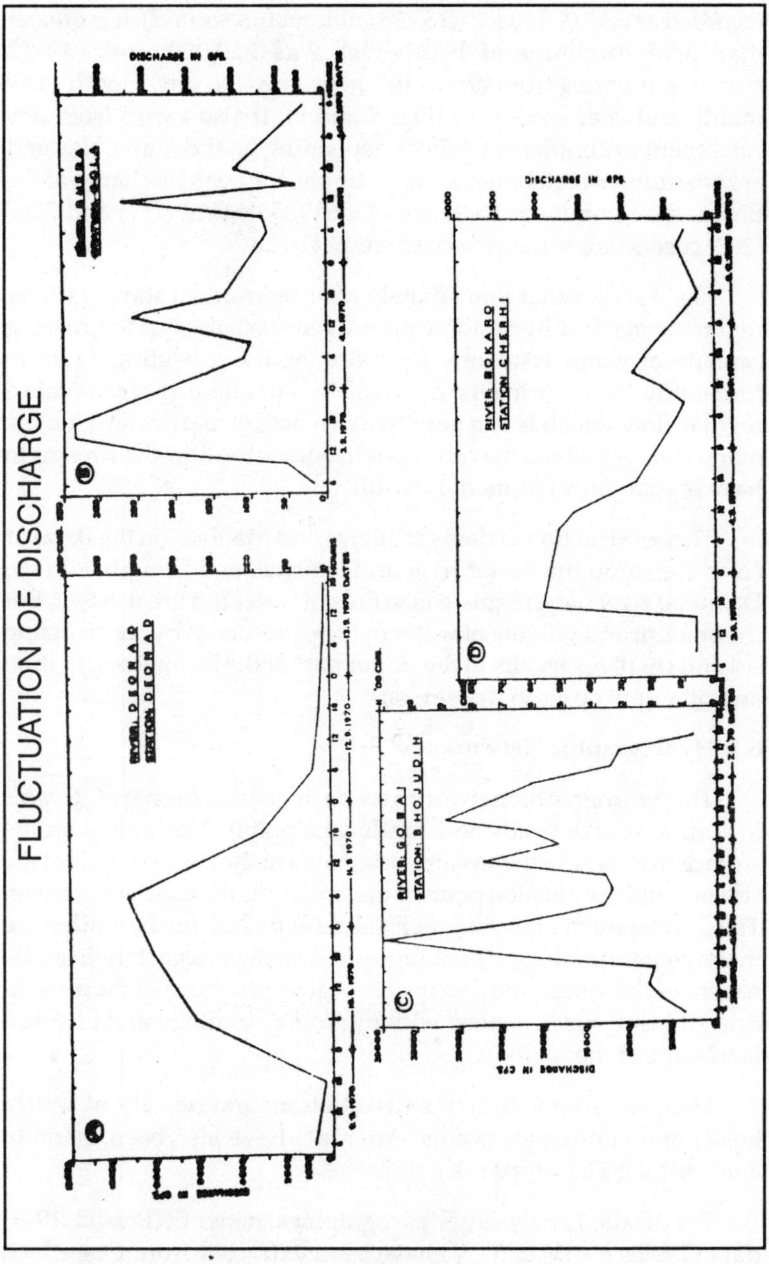
FLUCTUATION OF DISCHARGE
DISCHARGE IN CFS
RIVER: BOKARO
STATION: CHERHI
DISCHARGE IN CFS
DISCHARGE IN CFS

Fig. 9

recorded as 6,50,000 cusecs (18,408 cubic metres second). It is greater than peak discharge of 1943 which was 3,50,000 cusecs (9912 cumecs). It varies from hour - to - hour, day - to day, month - to - month and year - to - year (Figs.8 and 9). It also varies from one catchment to another catchment depending on the shape, size and area of different catchments. For example, the peak discharge of the Bhera, the Ijri and the Barki was 12563 (356), 4142 (117) and 7067 (200) cusecs (cubic metre second) respectively.

Fig. 9 shows that immediately after monsoonal showers river regime is marked by sudden short lived freshets. But the roaring cascade of water lasts only for a few hours, subsiding again to fordability. Even during the monsoon the streams do not maintain a regular flow which is of a very freakish nature, particularly on the marked slopes where the run - off is too quick to allow the streams to have regular flow (Ahmad, 1965:40).

The construction of dams at Tilaiya and Maithon on the Barakar river, Konar on the Konar river and Panchet, and Tenughat on the Damodar river have regulated the flow of water to a great extent. But the amount and volume of water may be preserved by constructing several small reservoirs in the upper part of the Basin for irrigation and other purposes in dry season.

6.4. Hydrographic Network :

The hydrographic network may be defined as the net of streams including source heads and confluence points. The points on the surface from where the streams originate are the source heads of the streams and the junction point of the streams is the confluence point. The development of hydrographical network has much significance in the geomorphological evolution of a drainage basin. It reflects the nature of the landscape and the geological structure of the area. In brief, it brings us in contact with hydrology, geology and physical landscape of the region.

Here we are to study the distribution and density of source heads and confluence points. Attempts have also been made to analyse their geomorphic significance.

For detailed analysis of hydrographical network (Prasad, 1979) data (Tables 6 - IV & 6 - V) have been extracted from One - Inch Topographical sheets to prepare density maps and graphs .

Table 6 - IV
Density of Source Heads

No.of Cat.	No.of S.H. per Sq.mile (2.59 Km.^{2})	FREQUENCY			% of the Cumulative Frequency	Density zone	% of Density zone.
		Mile^2	Km.^{2}	% of Total area			
I.	0	731	1893.29	11.65	11.65	Low	45.26
II	1-5	2109	5462.31	33.61	45.26		
III	6-10	1859	4815.81	29.63	74.89		
IV	11-15	964	2496.76	15.36	90.25	Medium	50.92
V	16-20	372	963.48	5.93	96.18		
VI	21-25	147	380.73	2.35	98.53		
VII	26-30	63	163.13				
VIII	31-35	15	38.85				
IX	36-40	8	20.72	1.47	100.00	High	3.82
X	41-45	4	10.36				
XI	Over - 45	2	5.18				
Total		6274	16249.66	100.00			100.00

Cat = Categories, S.H. = source Heads.

Table 6 - V
Density of Confluence Points

No.of Cat.	No.of C.P. per Sq.mile (2.59 Km.)	FREQUENCY Sq. Mile	Sq. Km.	% of Total area	% of the Cumulative Frequency	Density zone	% of Density zone.
I.	00	860	2227.40	13.71	13.71	Low	46.97
II	1-5	2087	5405.33	33.26	46.97		
III	6-10	1933	5066.47	30.81	77.78		
IV	11-15	813	2105.67	12.96	90.74	Medium	49.21
V	16-20	341	883.19	5.44	96.18		
VI	21-25	175	453.25	2.78	98.96		
VII	26-30	46	119.14				
VIII	31-35	11	28.49				
IX	36-40	5	12.95	1.04	100.00	High	3.82
X	41-45	2	5.18				
XI	over 45	1	2.59				
Total	6274	16249.66	100.00				100.00

Cat. = Categories, C.P. = Confluence points.

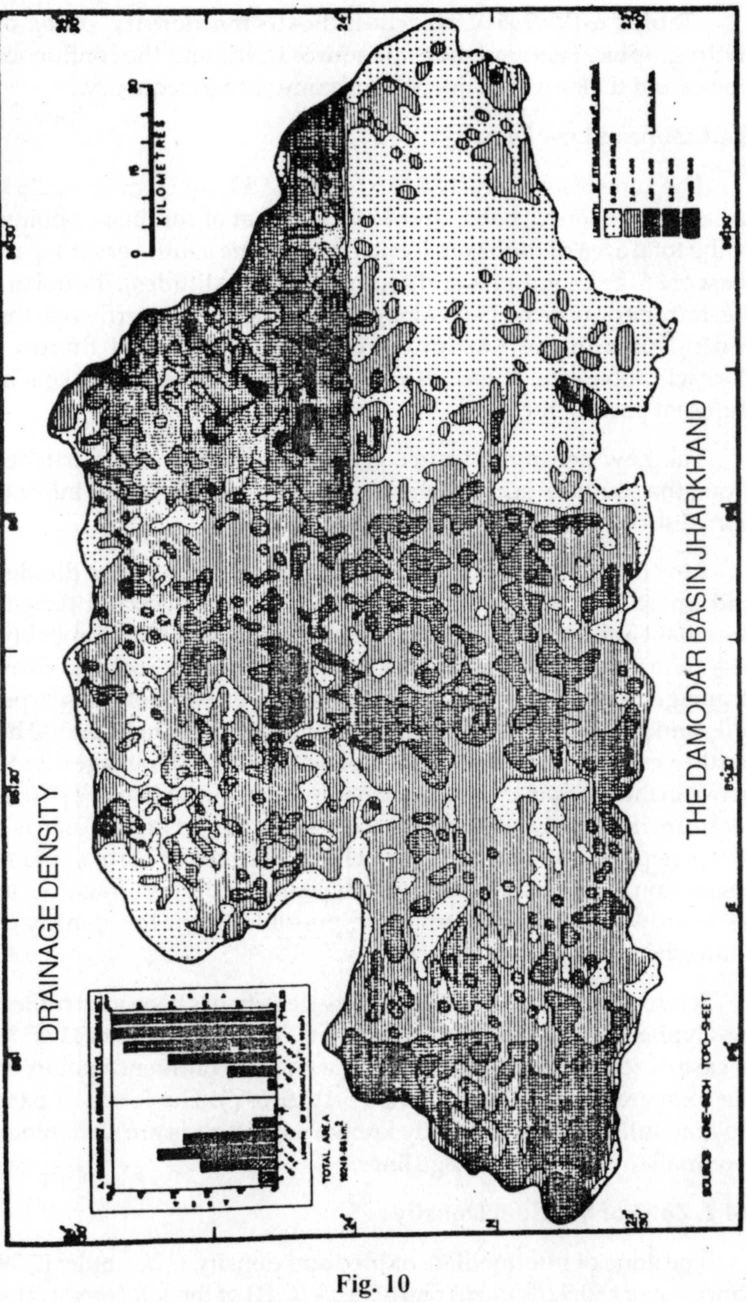
DRAINAGE DENSITY
THE DAMODAR BASIN JHARKHAND
KILOMETRES
TOTAL AREA
TOPO-SHEET

Fig. 10

Tables 6-IV & 6-V clearly indicate the density zones of hydrographical networks, i.e. the source heads and the confluence points and their areal frequency and cumulative frequency.

6.4.1. Zone of Low Density :

The zone of low density 0 to 5/ mile^2 (2.59 Km.^{2}) covers 45.26% in case of the source heads and 46.97% as that of confluence points of the total area of the basin. It extends over the south - eastern part (east of 86^0 E. Longitudes and south of 24^0 N. Latitudes), the north - western part, narrow belts along the divides and interfluves, the central part in the Karnpura coalfield and around Bundu - Barkachumba area. It has also been observed in small pockets in different parts of the Basin.

The Low density of source heads and confluences in patches along the southern scarp of the Basin is due to high rate of infiltration of surface water in crystalline lime - stones.

Low density of source heads and confluences along the divides and interfluves is due to 'crictical area' where overland flow is dominant and erosive power (energy) of surface water is, probably, less than the resistance of soils and exposed bed rocks to form drainage lines. This is why most of the upper parts of the scarps, hills and ridges have low density of hydrographic networks. The north-western part of the Basin is a plateau surface. Drainage divide between the Barakar and the Mohana also runs through it. The parent rocks are invariably jointed. The south-eastern part of the Basin is a low peneplain dotted with isolated hills. Absence of scarps, sparse vegetation, presence of coarse sandy loam, faults and fractures in the Gondwana areas are perhaps responsible for the low density of source heads and confluence points.

It is to be noted that the zone of low density has been interrupted by 'Cypher Zone 'which has no streams. This Zone covers 11.65 % in case of source heads and 13.71 % in case of confluence points of the total area of the Basin (Table : 6-IV & 6-V). The forms of bare isolated hills and knolls locally known as Dungaris are sometimes too small to develop drainage lines.

6.4.2. Zone of Medium Density :

The zone of intermediate or medium density, 6-20/ mile^2 (2.59 Km^2), covers 50.92 % (S.H.) and 49.92% (C.P.) of the total area of the

TABLE : 6-VI
Drainage Density

No of Cat.	Total length mile/sq.mile (2.59 Kms.2	FREQUENCY sq. Mile	sq. Kms.	% of total area	% of Cumu-lative Frequency	Density zone zone	% of Density
I	0	348	901.32	5.55	5.55	LOW	28.55
II	.01-2	1443	3737.37	23.00	28.55		
III	.2.01-4	2248	5822.32	35.83	64.38	MEDIUM	63.85
IV	4.01-6	1758	4553.22	28.02	92.40		
V	6.01-8	453	1173.27	7.22	99.62	HIGH	7.22
VI	8.01.10	22	56.98	0.35	99.97	VERY	0.38
VII	Over - 10	2	5.18	0.03	100.00	HIGH	
TOTAL :-		6274	16249.66	100.00			100.00

Source : One - Inch Topographical Sheets.

TABLE : 6-VII
Stream Frequency

No. of Cat.	No. of Streams per Sq. mile (2.59 Kms.2)	FREQUENCY Sq. mile	Sq. Kms.	% of total area	% of Cumulative Frequency.	Texture zone	% of Texture zone.
I	0	348	901.32	5.55	5.55	COARSE	25.08
II	1-5	1226	3175.34	19.53	25.08		
III	6-10	1891	4897.69	30.14	55.22		
IV	11-15	1521	3939.39	24.24	79.46	MEDIUM	64.72
V	16-20	649	1680.91	10.34	89.80		
VI	21-25	314	813.26	5.00	94.80		
VII	26-30	251	650.09	4.00	98.80	FINE	9.77
VIII	31-35	48	124.32	0.77	99.57		
IX	36-40	17	44.03	0.28	99.85		
X	41-45	6	15.54	0.10	99.95	ULTRA	-0.43
XI	46-50	2	5.18			FINE	
XII	Over 50	1	2.59	0.05	100		
Total -		6274	16249.60	100.00			100.00

Source : One - Inch Topographical Sheets.

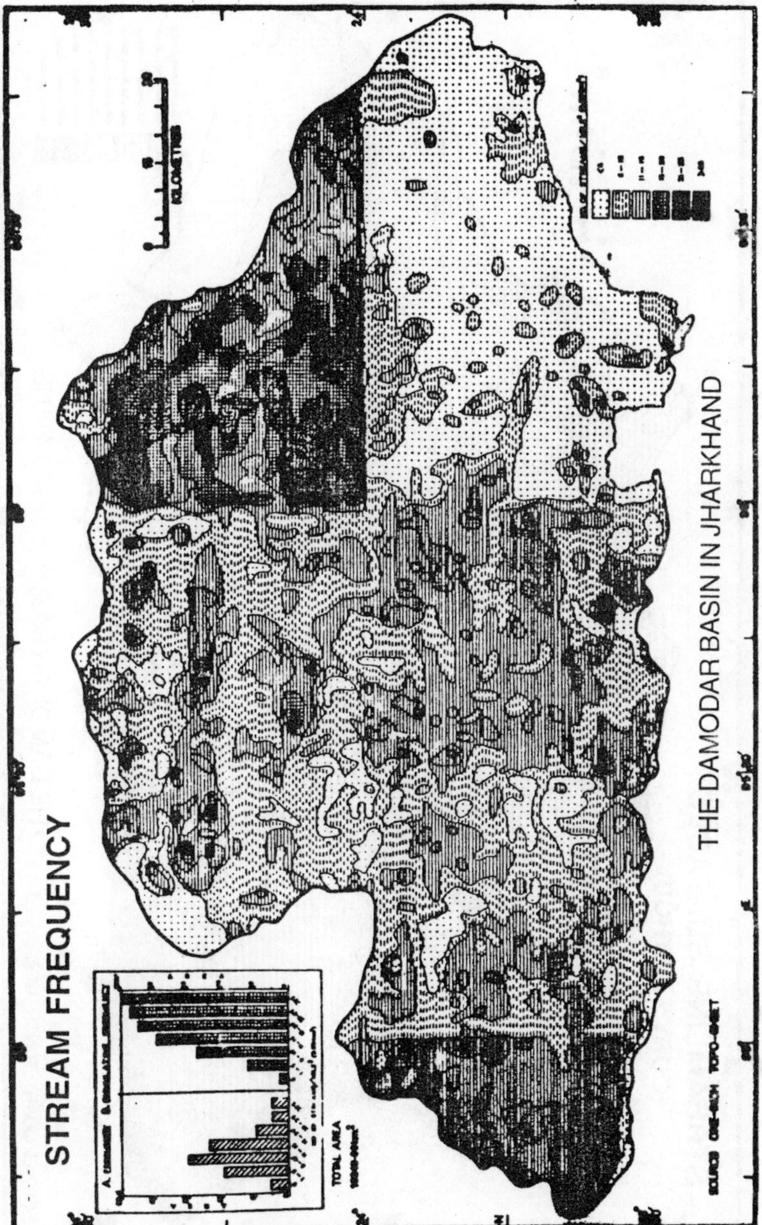
STREAM FREQUENCY
THE DAMODAR BASIN IN JHARKHAND
KILOMETRES
TOTAL AREA

Fig. 11

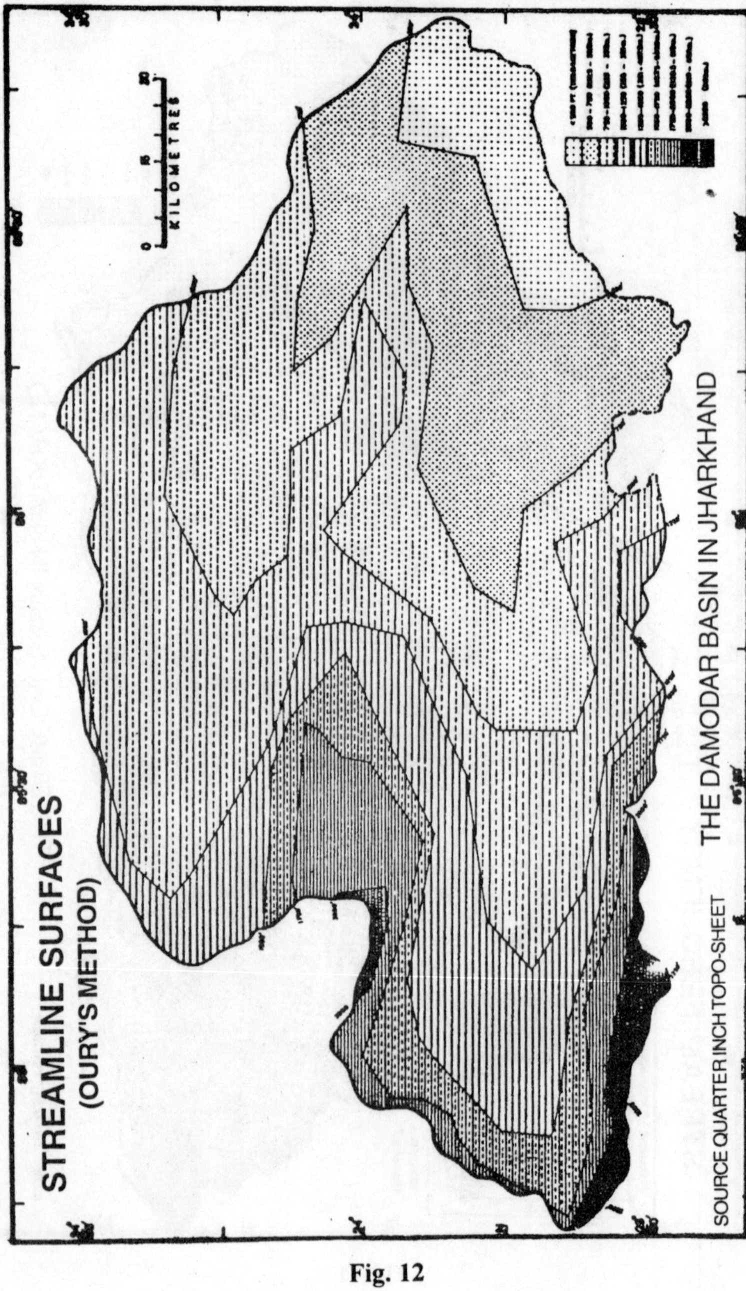
STREAMLINE SURFACES
(OURY'S METHOD)
THE DAMODAR BASIN IN JHARKHAND
SOURCE QUARTER INCH TOPO-SHEET
KILOMETRES

Fig. 12

TABLE : 6 - VIII

Stream - Line Surfaces :

ELEVATION In feet.	In metre	Sq. Mile	FREQUENCY Sq. Kms.	% of total area.	% of Cumulative Frequency.
500	152	369.36	956.65	58.88	5.88
500-750	152-228	1066.16	2761.35	17.00	22.88
750-1000	228-305	1140.12	2952.92	18.17	41.05
1000-1250	305-381	1741.02	4509.24	27.75	68.80
1250-1500	381-457	936.04	2424.34	14.92	83.72
1500-1750	457-534	448.72	1162.18	7.16	90.88
1750-2000	534-610	435.46	1127.84	6.94	97.82
Over 2000	610-686	137.12	355.14	2.18	100.00
Total -		6274	16249.66	100.00	

TABLE : 6 - IX
Physiography

ELEVATION		FREQUENCY			% of Cumulative
Feet	Metres	Sq. mile	Sq. Kms.	% of the total area	frequency.
500	152.5	139.90	362.34	2.23	2.23
500-750	152.5-288	701.5	1216.88	11.18	13.41
750-1000	228-305	931.10	2411.55	14.85	28.26
1000-1250	305-381	1993.44	5163.00	31.77	60.03
1250-1500	381-457	1078.48	2793.24	17.19	77.22
1500-1750	457-534	512.44	1327.22	8.16	85.38
1750-2000	534-610	572.04	1481.57	9.12	94.50
2000-2250	610-686	305.12			
2250-2500	686-762	25.60	893.86	5.50	100.00
2500-2750	762-839	6.56			
2750-3000	839-915	3.36			
Over-3000	915		4.48		
Total			6274	16249.66	100.00

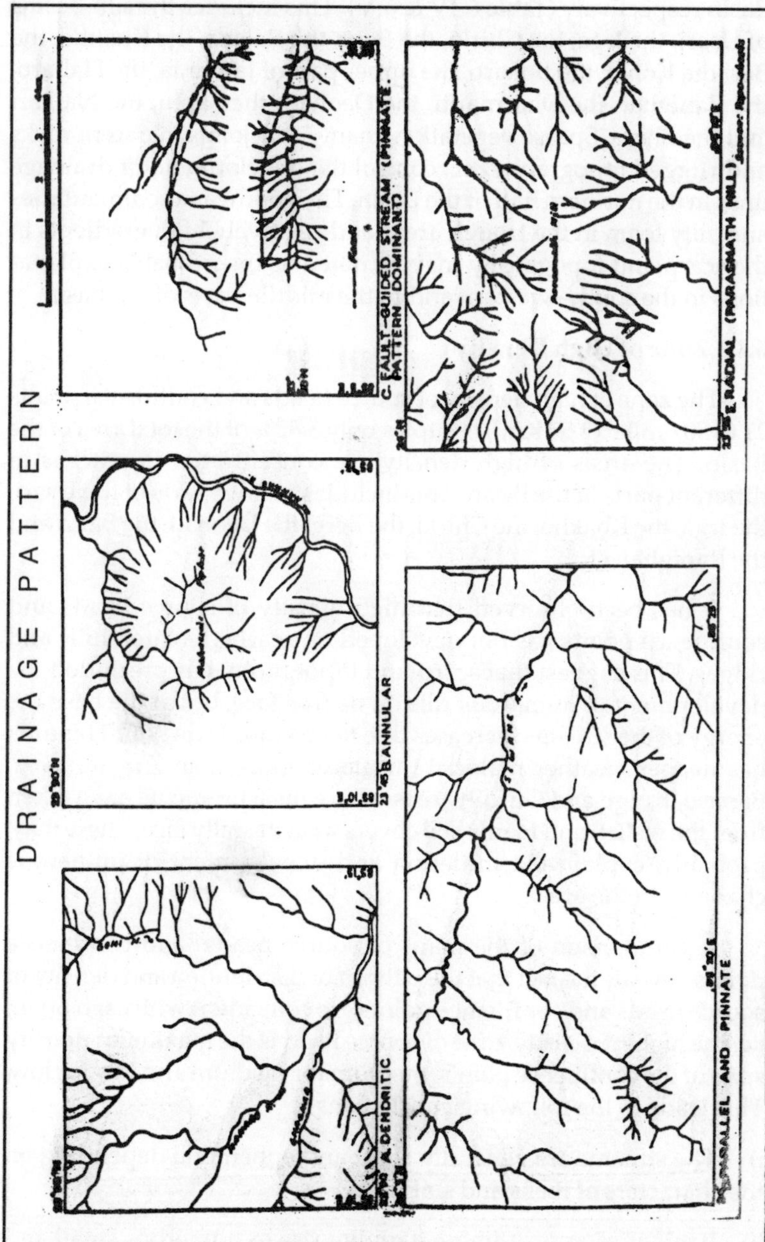
DRAINAGE PATTERN
A. DENDRITIC
B. ANNULAR
C. FAULT-GUIDED STREAM (PINNATE PATTERN DOMINANT)
PARALLEL AND PINNATE
E. RADIAL (PARASNATH HILL)

Fig. 13

Basin respectively (Table 6-IV & 6-V). This includes the sub-basins of Usari, the Irga, the Chirki, the Berg, the Rajoya, the Khakho, the Beri the Konar, the Bokaro, the upper part of Jamunia, the Haharo, the Ranighat, the Dharmauti, the Deonad, the Saphi, the Nalkari and the Bhera. Sparse vegetation, loamy soil, joints in parent rocks and rugged topography etc. control the development of drainage lines in the northern half of the Basin. The gondwana sedimantaries and silty loam in the trough areas and relatively high gradients in the scarpland topography may be noted as the probable explanations in the south - western and in the middle parts of the Basin.

6.4.3. Zone of High Density :

The zone of high density of source heads and confluence points 21 to 50/ mile^2 (2.59 Km^2) occupies only 3.82% of the total area of the Basin. The areas of high density are concentrated in patches in different parts of the Basin. This includes mainly parts of the Usari, the Irga, the Khakho, the Chirki, the Berg, the Deonad, the Barki and the Ranighat, etc.

It has been observed that high density of source heads and confluence points has not developed on scarps, isolated hills and ridges. This suggest that scarpland topography has precluded the development of numerous rills on its free face, but at the base the energy of the streams increases due to increased run - off. This area has deeper weather material (at places more than 7 meters, e.g. Banaso, Kulgo and Giridih areas where measurements have taken from the wells)) and loamy soil covers than the hilly face. These may, probably, explain the initiation and development of numerous channels in the area.

A comparison of the maps of source heads and confluence points reveals the fact that the pattern of distribution and density of source heads and confluence points are coincident with each other, i.e. the highest density zone of source head is the maximum density zone of the confluence points, medium for medium and low for low. This leads to the following conclusions :

i) The streams are generally short and ephemeral depending on the characters of rocks and soil cover.

ii) Bank erosion is prominent giving rise to numerous small and parallel streams, e.g. in the Chirki sub-catchment.

iii) Joints, faults and fractures in the rocks have greater control over the initiation and development of the drainage lines.

The above discussions conclude that the pattern of the distribution and density of hydrographic network is largely controlled by the nature and character of rocks, geological structure and soil cover, etc. They are not related to the gradients of the area as high gradient has low density, etc.

6.5. Drainage Texture:

The texture of drainage may be defined as "relative spacing of drainage lines implying the drainage density and stream frequency" (Ahmad, 1965 : 19). It may be graded as " Coarse, medium, fine and ultrafine, depending on the degree and extent of dissection of topography by streams" (Sinha, 1968 : 45). Horton (1945) has pointed out that what we commonly refer to as the texture of drainage really includes both drainage density and stream frequency, Melton (1966) analysed in detailed the relationships between drainage density and stream frequency, both of which measure the texture of the drainage net, but each of which treats a separate aspects. Drainage density is defined as " the total length of streams per unit area" (Gresswell, 1972 : 178), while stream frequency means " the number of streams divided by the area of basin" (Ahmad, 1965b : 19).

For an analysis of the texture of drainage of the Damoder Basin in Jharkhand stream lengths have been measured per square mile (2.59 $km.^2$) and number of streams has been counted per grid of 1 $mile^2$ (2.59 $km.^2$). Data (Tables : 6-VI and 6-VII) have been recorded at the centre of each grid. Maps (Figs. 10 and 11) and histograms (Figs. 10A, 10B, 11A and 11B) have been prepared to mark the seperate zones and to show areal extent respectively.

In this regard it is essential to note that the texture of drainage is always approximately true, because " area and length are measured from a map, which projects the sloping surface and channels upon a horizontal plane, the measured values are somewhat less than the true values, to a degree depending upon the amount of slope "(Strahler, 1971 : 491).

Figs. 10 & 11 and Tables 6 - VI and 6- VII reveal that the texture of drainage is coarse, medium, fine and ultra fine in the Damodar Basin in Jharkhand.

6.5.1. Coarse Texture :

The zone of coarse texture of drainage (density 0 - 2 miles/ mile^2 or 0-3.2 Kms per grid of 2.59 km.2 and stream freauency 0 - 5/mile^2 or per grid of 2.59 Km.^{2}) occupies 28.55 - 25.08% of the total area of the Basin. It extends over the south - eastern part, the north - western part, narrow belts along the northern and the southern divides, the interfluve areas around 24^0 N.Lat. and 85^{0}30' E. Long. the central part of the Karnpura coalfield and around Bundu and Barkachumba areas of the Damodar Basin.

The coarse texture around Bundu - Barkachumba area is due to high permeability of crystalline limestone. Narrow belts of coarse texture along the divide is probably due to ' Critical ' area where the erosive power of moving water is less than the resistance of soil and exposed bed rocks to form the drainage lines. This critical area may be said as the zone of overland flow. Similar is the case with the interfluvial zones between the Siwani and the Konar and also between the tributaries of the Barakar and the Siwani and the Mohna, etc. But, in these areas the streams are advancing towards the crest of the divides by their head - ward erosion. The north - western part of the Basin is a flat plateau and drainage divide also runs beside it. The rocks are invariably jointed. These factors are probably responsible for the coarse texture of drainage. The south - eastern part of the Basin, east of 86^{0}E.Long. and south of 24^0 N. Latitudes, is one of the four erosion surfaces of the Basin. Absence of scraps and vegetation in the Gondwana basins perhaps are responsible for the coarse texture of drainage.

It is interesting to note that the zone of coarse texture of drainage has been interrupted by patches of cypher area which has no streams. It occupies 5.55% of the total area of the Basin. The bare isolated hills and knolls, etc. are generally too small to develop any discernible drainage lines.

6.5.2. Medium Texture :

The zone of medium texture covers a vast area (63.85 - 64.72%) of the Basin (Table 6 - VI & 6 - VII). It occurs partly or wholly in the sub - basin of the Ambagarha, the Bhera, the Khanjo, the Konar, the Bokaro, the Saphi, the Nalkari, the Dharmauti, the Ranighat, the Barki, the Haharo, the upper Jamunia, the Usari, the Irga, the Chirki, the Berg, the Khakho, the Beri and the Rajoya. This texture zone has drainage density of 2.01 to 6.00 miles/mile^2 or 1.25 to 3.72 Km./Km^2 and stream frequency 6 to 20/mile^2 (2.59 Km.^{2}).

The general absence of vegetation and relatively thick loamy soil may be the probable explanation for the medium texture of drainage in the northern half of the basin. In the southern part of the Basin the presence of loamy residual soil, Gondwana sedimentaries, dense forest, high gradients, etc. may be the probable explanations for the medium texture of drainage.

6.5.3. Fine Texture :

The zone of fine texture of drainage occupies 7.22 - 9.77% of the total area of the Basin. It has drainage density of 6.1 to 8 miles/mile^2 (2.59 Km^2) and stream frequency 21 to 35 / mile2 (2.59 Km.^{2}). It is seen in patches in the Usari, the Khakho, the Irga, the Deonad, the Ranighat, the Barki, the Haharo, the Konar, the Bokaro and the Barsoti sub - basins.

6.5.4. Ultrafine Texture :

The ultrafine texture of drainage covers a very little area (0.38 - 0.43%) of the Basin. It has been seen in small pockets in the Deonad, the Ranighat, the Barki, the Irga and the Usari sub - basins. This zone has drainage density 8 to 12 miles/mile^2 (2.59 Km.^{2}) and stream frequency of 35 to 55 mile^2 (2.59 Km.^{2}).

In the Usari and the Khakho areas, it includes generally impermeable shale characterised with relatively sparse vegetation. In the upper reaches of the Barsoti sub - basin, deep residual soil, reflected by steep side slopes of the hills, gives rise to a pocket of fine texture of drainage. The texture and pattern of drainage have been, probably, developed and controlled by the foliation of structure of the parent material. The weak materials are more easily eroded to form valleys and more resistant materials form ridges.

In most of the areas water rushes down the scarpland topography, the hills and the ridges give rise to the fine texture of drainage.

The above discussions conclude that the texture of drainage is not related to the difference in the relief of the area. We see coarse to medium texture in the hills and scarpzones as well as in the plantation surface areas. most of the isolated hills are cypher zones. Fine and ultrafine texture of drainage may be seen in areas of shale, silty soil and near the scarp zones and hills.

They are most probably guided and controlled by the nature of rocks, the nature and characters of soils, faults and fractures in the Gondwana Sedimentaries, invariable joints in the parent rocks, climate and vegetation, etc. in the Damodar Basin in Jharkhand.

6.6. Stream - Line Surface :

The stream - line surface (Prasad & Gupta, 1979) is that on which the available relief is zero. It is defined by contour passed through the intersections of actual contours with principal streams (Dury, 1952 : 140). Thus, it indicates the amount of denudation to be achieved (Mokhouse and Wilkinson, 1976 : 66). Each line outlines a class correlation with existing planation surface of the Damodar Basin in Jharkhand.

The stream line surface map (Fig. 12) of the Damodar Basin in Jharkhand has been prepared by plotting intersections of actual contours within the main streams. The area of each class has been measured (Table 6 - VIII) and frequency and cumulative frequency have been calculated.

Fig. 12 and Table 6 - VIII reveal the following facts :

i) The comparison between the two maps - physiography (Fig. 3) and stream - line surface (Fig. 12) clearly shows the areas, deeply dissected by rivers and extensive interfluves surviving in between the valleys.

ii) The stream - line surface mark somewhat a new physiographic zone, different from the present relief features. These have extended the probable lower areas and minimised the upper surface.

iii) The rivers of the Basin are active in denudation of the area. Rapid erosion is associated with gorges and narrow valleys, intensity of run-off, amount of load, discharge channel gradients and ruggedness of the surface, etc.

iv) It gives us an idea of the direction of denudation. It seems that scarp - zones are receding rapidly after the uplift of the area during the Himalayam Orogeny, which has given a new life to the streams of the area. This is also evidenced by the comparison between the Table 6 - VIII & 6 - IX of stream - line surface and physiography respectively.

v) The Bhera, the Dhabdhab, the Nalkari, the Saphi, the Barki, the Hahara, the Barakar, the Kewta and the Barsoti rivers in the scarp zones are very active resulting in the apparent scarpretreat. The hills in front of scarps may be taken as proof that the scarps have been retreating. These hills have, probably, been left isolated by retreat of the scarpzones in course of active erosion by the streams. An excellent example may be observed along the southern scarp of the trough. Rochap hill (655 m.) has been left isolated by the retreat of the scarp in course of active erosion by the tributaries of the Nalkari and the Batuka rivers. The hills between the Gorha nala and the Khero river may be another notable example. It has been left isolated by retreat of the scarp - zone in course of active erosion by the tributaries of the Barsoti.

vi) The Khudia, the Chirki and the Katri rivers with their tributaries are engaged in narrowing the Tundi hills. If this would have been of softer rocks, the stream line surface of 228 metres and 305 metres would have receded upto the Parasnath hills. These hills owe their preservation to their hardness. Inspite of their hardness these rivers will probably isolated these hills in course of time and then they shall remain as mere monadnock on the lower eroded surface.

vii) There is no doubt that the Damodar Valley project has slackened the intensity of erosion and retreat of scarps to some extent, but it is a natural phenomenon in the Basin which prevailed in the past,is active at present and will characterise denudation in future.

The above discussions conclude that the Damodar Basin in Bihar is under rapid denudation by numerous active streams. However, a greater amount of denudation is still to be achieved.

6.7. Drainage Pattern :

Drainage has an important place in the geomorphological study of any particular area of country as it is mainly guided by the structure, the amount of the slope and the nature of the topography. Streams are potent agents of erosion. They are always active to change the face of the landforms. "Drainage refers to the manner in which precipitation falling within an area or brought from outside is drained off" (Ahmad,E.. 1965 : 27). In the Damodar Basin in

Jharkhand the rivers are based on precipitation. In rainy season they get more water and become turbulent and in summer season most of them are dry.

It is important to note that " rivers, with their tributary - system, are the main channels of drainage of the land - surface; they are at the same time also the chief agents of land - erosion and sculpture and the main lines of the transport of the products of the waste of the land to the sea" (Wadia, 1975 : 24). So, we see that rivers serve to reduce the height at one place by erosion and cause new features by deposition. The Damodar river along with its tributaries acts as an erosional agent in the source region and a depositing agency in the plain of west Bengal.

The following characteristics of the rivers of the Damodar Basin may be observed :

(1) The streams pass through narrow, rocky and steep sided valleys (gorges). They form beautiful rapids and waterfalls. The Chandro waterfall at the konar river is one of the notabe examples. We see here a fine example of the recession of waterfall. Above gorge, an example of terrace may also be observed.

(2) The Damodar is a fault - guided river. This can be clearly observed above the confluence of the Haharo to the Damodar. Here the Damodar follows the fault line. The Chundru Nala, a tributary of the Barki also follows the fault line.

(3) The breaks in the longitudinal profiles of the streams are rejuvenation heads. The rejuvenation head at Bermo where Konar dam has been constructed may be observed as one of the beautiful examples. Another example of rejuvenation head is at the Usari waterfall on the Usari river. The gradient above the head of the fall is 1 : 900 while in the lower part it is 1 : 300 or 400. This clearly indicates the break in slope (Bose, S.C., 1945).

(4) The velocity of water is relatively high due to high gradients of the channels in the scarpland topography.

(5) Most of the rivers are seasonal in character. They do not receive water during dry summer and water level goes beneath the river beds. Even during rainy season they do not get regular supply of water due to uncertain monsoon and rugged topography in the

source region.

Pattern :

Drainage pattern means :the layout of the rivers and density or infrequency . . streams per unit area" 'Ahmad (1965a : 32). It also "refers to particular plan or design which the individual stream courses collectively form" (Thornbury, 1954 : 120). Feldman, et al. (1968 : 284) defined drainage patterns or arrangements" as spatial relationship among streams or rivers, which may be influenced in their erosion by inequalities of slope, rock resistance, structure and geologic history". Its study presents one of the most practical approaches to an understanding of structural and lithological control of landscape evolution.

Broadly speaking, the common drainage pattern in the area is what we call dendritic (13A), first used by Russel (1898) and later defined by Cleland (1916) as " non - systematic" or like a tree, "having stem branches limbs and twigs" (Von Engeln, 1948 : 138). In the area under study we do not find all parts of the dendritic pattern as distributaries are found in the delta region of west Bengal and the Damodar Basin in Jharkhand is away from that area. So, no third part, which is compared to the roots of the tree, is found in the area under study. In the Basin we see only branches i.e. tributaries and the trunk line rivers lie the trunk of the tree.

Nature allows trees to develop their branches and similar is the case with the rivers. The rocks of the plateau areas and trough belt are of varying character. Joints and faults are also there to control the pattern of the drainage. It has been noted earlier that the area was uplifted epeirogenetically during the Himalayan orogeny. Therefore, the rivers were also subject to rejuvenation.

Incised meanders, scarp waterfalls, terraces, breaks in the long profiles of the rivers, etc. are clues to favour the idea of the successive uplift of the Basin. Hence, we may conclude that the drainage in the Basin is "modified dendritic pattern".

It is clear from the fore-going that the dendritic pattern of drainage predominates in the region. But we do not ignore the importance of radial pattern (Fig.13A) which is "short and radial rising from the highlands, tumbling headlong down the marginal scarps" (Spate, 1954 : 753). So, the slopes of the hills and highlands are the guiding factors for the development of such pattern of the drainage. The draiange pattern of the Parasnath hill areas, Baragaon Pahar, Chainpur hills, Lugu Pahar, Jhumara hill and Mahabar hill, etc. are some of the notable examples of radial pattern.

An annular or ring like pattern (Fig. 13B) may be observed around the Mahudi hill. The hill and its surroundings have an alternating belts of hard (sandstones) and soft (shales) rocks.

The pinnate pattern of drainage is "feather like" (Ahmad, 1965b : 17) with tributaries parallel and joining the master stream at acute angles (Feldman, et . al. 1968 : 287). This is a phenomenon of the faulted basins. The Damodar is doubtless a faulted basin, where the tributaries from the scarps adjoining the trough courses giving a pinnate pattern (Figs. 13C and 13D).

Structural controls, mainly joints in granites and gneisses and faults in the trough area, have led to regular spacing of streams. parallel or near - parallel, forming parallel pattern (Fig. 13D). The tributaries of the Siwani form the parallel pattern. A glance on the map reveals numerous examples of parallel pattern of drainage in the Basin.

A series of parallel faults has resulted in alternating bands of hard and soft rocks which gives the drainage " a fault trellis " pattern. This pattern can well be observed in the Karnpura basin.

The above discussions conclude that the spatial relation of the drainage has been controlled by the structure, slopes, topography, earth movements and geomorphological history of the area.

6.8. Drainage Divide :

Between one river basin and the next, an area of higher ground divides the water draining into one river from that draining into its neighbour. The line (Cain, 1976 : 56) along the top of this high ground is the 'drainage divide' or 'water shed' or 'water parting'. There are minor watersheds between all the various streams and tributaries inside the Basin.

Drainage divide plays a major role in the development of streams. It also controls the size of the catchment and the flow of water in the Basin. Davis (1909 : 26) has rightly remarked that "there is no more beautiful process to be found in the systematic advance of a geographical cycle than the definition, subdivision and rearrangment of the divides (water - partings) by which the major and minor drainage basins are separated" (Davis, 1909:26). It has been noted earlier that the Damodar Basin in Bihar has been delimited by a well - marked divide. The Basin perimeter runs along the line of divide which separates the Basin from the south Koel and the Subarnrekha in the south, the Auranga and the Mohna in the west, the tributaries of the Ganges in the north and the Ajoy in the north - east.

The study shows clearly the differences in the elevation of the divide. If one walks along the surrounding divides of the Basin starting from Khamarpat (84^{0}41' : 23^{0}37' N/ 1068 m.) in the south - west to Chauparan (400 m.) in the north - west via Chandwa-Tori Balumath ridge (600 m.), Kasiatu Hill (676 m.), Hedlag (600 m.), Mahuda (676 m.), Itkhori (414 m.), from Chauparan to Deori (380m)in the north east via Domchach (411m), Deori to Salampur (152 m.) in the east via Maheshmunda (327 m.), from Salampur to Bhaski in the south - east following the state boundary; and from Bhaski to the starting point Khamarpat via Sagrampur Pahar (549 m.), Bariatu (630 m.), Baragaon Pahar (1038 m.) Pithouria hill (794 m.), Mandar (700 m.), Nakta hill (804 m.), Barkichanpi (681 m.), following the southern divide, the differences in elevations of divide are well observed.

The differential erosion and river capture are changing the line of divides. The migration of divide results in contraction and expansion of the Basin area. This has been clearly observed at several points. In the south - eastern parts of the Bhera sub-catchment (24^{0}31' : 85^0 48'), a tributary of the Subarnrekha river (out side the Basin under study) by capturing the head stream of a tributary of the Gomti river, has pushed the divide north. The tributary originates from the western face of Segrampur pahar (549m.). There is no topographic eminence to divert the stream to the south. An elbow of capture at the railway cutting has also observed. By capturing the stream, some of the Basin areas have also been captured.

South of Kasiatu hill near Sibla village the source streams of the Barki river pushed the divide westward and expanded the Basin area. The source streams of the Gaghari, a tributary of the Auranga, Originate on the eastern face of the Latdag hill and after flowing east and south - east for a short distance, they turn towards west through

Tori gap (480 m.) here, it is peculiar that the divide between the Auranga and the Damodar does not run through the highest line of elevation.

On the north-eastern part of the Basin a clear case of migration of divide is observed between the Usari and the Painti, a tributary of the Ajoy river.

The migration of divides by headward erosion may be observed within the basin. It is most probable that the Siwani has diverted the water of most of the source streams of the Barakar. The source streams of the Jamunia are also pushing back the divide by headward erosion.

The primary, the secondary and the tertiary divides are observed within the Basin.

The primary divide separates the Damodar from its own tributary the Barakar. The line of divide runs west - east through Bagodar upland, on which lofty igneous intrusive (Parasnath hill, 1366 m.) stands like a pillar. The important river sub - catchments have been separated by the secondary divides. The tertiary divides separate several small basins and sub-basins.

The construction of dams and reservoirs in the Basin has produced a new scheme to saparate dam catchments and sub-catchments. Five major divides have produced six major Catchments. Out of these Khudia catchment is outside the dam - catchment areas. Within these major catchments, following the divides and interfluves, 39 sub-catchments have been separated. Out of these sub-catchments, Panchet dam (right), sub - catchment and parts of Maithon dam sub - catchment, Gowai sub - catchment and Ijari sub - catchment are within the state of West Bengal.

The Study of divides in the Basin reveals the following facts :-

(1) It is observed that the headward erosion of the streams has a close relation with the migration of divides. " By the recession of valley heads towards the crest, the evolving watershed grows progressively more sinuous" (Pippan, 1968 : 1223). For example the source streams of the Saphi are pushing the line of divide toward the gentler slope into the Ranchi planation surface.

(2) The migration of divides by headward erosion and river piracy may probably be explained by the fact that the area has been uplifted

epeirogenetically in the recent past. The process, still continuing in the form of tilts, may be responsible for the asymmetric gradients along two opposite competing rivers.

(3) The Basin has distinct divides. So, it is most probable that the Basin is not in the initial stage because in that stage the divide should be indistinct.

(4) During the stage of maturity "the divides between adjoining valleys and drainage basins become rounded crests that gradually lose height" (Holmes, 1975 : 473). In the old stage the area should be characterised with " faint relief". But we observe sufficient relief in the upper part of the Basin. So, it is just possible that the Basin is not in the old stage.

It is, probably, passing through youth to maturity. In other words we may say that youth and maturity paly hand in hand in the Basin.

7

Morphometric Analysis of the Relief Aspects

Relief is generally used in the same sense as the physical features but we know that its precise meaning is the difference between the highest and the lowest points of the particular area. The entire surface slopes in one direction or another. Thus the study of slopes and relief has ' nodal position ' in geomorphological research. The present chapter, therefore, includes the study of gradients and the relief measures of the Damodar Basin in Jharkhand.

7.1. Slopes:

"Slopes are ubiquitous elements of the landscape" (King, 1966 : 17). The entire landscape is made up of slope elements and thus geomorphology of the Damodar Basin in Jharkhand is largely concerned with the study of gradients.

7.1.1.Channel Gradients:

7.1.1.1. A Single Channel Profiles:

Channel slope is defined as " the ratio of vertical drop to horizontal distance, measured from the upper end to the lower end of the single stream segment" (Strahler, 1971 : 492). For thalwegs the stream lengths have been measured with the help of rotameter. The vertical drops of the streams have been obtained from the topographical sheets. The thalwegs or valley way profiles of the Damodar river in Jharkhand and the Barakar river have been constructed taking suitable vertical and horizontal scales .

The valley way profiles of the rivers reveal the following facts:

(i) The general forms of the thalwegs are concave upwards, but the degree of concavity varies considerably from river to river.

(ii) The regularity of curves varies with different rivers, some being concave only if the irregularities are ignored.

(iii) The thalwegs of the Konar, the Bhera and the Bokaro are generally concave, occasionally seperated by convex sections.

(iv) The thalwegs reveal break in slopes, probably representing the rejuvenation heads or 'knick - points'.

(v) The Damodar shows graded profile in the lower sections (gradient 1 in 1000) i.e. the character of maturity. But on the upper section, high gradient (1 in 5 or 6) is probably the feature of youth stage.

7.1.1.2. Composite Profile :

"A composite stream profile combines the stream segment of the given order within the watershed into a single average segment whose vertical drop is the mean drop of all the segment and whose distance is the mean horizontal lengths of all segment" (Strahler, 1968 : 905).

For composite stream profiles slopes of all channel segments of the first order have been obtained and averaged. The mean slope has been denoted by Si. The mean lengths (Lu) of the streams have also been obtained for the same order. Similarly data for other order streams, i.e. second, third and so on, have been obtained, slope ratio (Rs) has been calculated for them (Table : 7 - I). Data have been plotted in Fig. 14.

TABLE : 7 - I
Channel Gradients

Order U	The average vertical drop HU in metres.	Mean length LU in metres.	Mean channel slopes SU *	Slope ratio Rs**
1st.	67.50	2700	0.025	0.468
2nd.	39.78	3400	0.0117	
3rd.	40.98	6830	0.006	0.5128
4th.	95.78	18420	0.0052	0.866
5th.	79.48	30570	0.0026	0.5
6th.	97.35	64900	0.0015	0.5769
7th.	109.43	109430	0.001	0.6

Average Rs=0.589

*S *SU = Hu/Lu, Where = Su = Mean Channel Slope of order U; Hu = The average vertical drop of order U; Lu = Mean stream length of order U : Rs = slope ratio. ** Rs= Su/Su-1

The study reveals the following facts :

(i) The mean slope is decreasing with the increase in the value of the order of the channel segment. It varies from 0.025 in the first order to 0.001 in the seventh order.

(ii) The logarithms do not show perfect alignment with straight line. Points of second, third, and fifth orders are off the straight line. But they fall near it and hence we can say that the agreement is good to a great extent.

(iii) The average slope ratio (Rs) is 0.589. But it ranges from 0.468 to 0.866. We see the highest slope ratio (Rs) 0.866 between the third and the fourth orders. These unusual changes in slope ratio (Rs) cannot be explained byclimatic variations. The differential erosion partly caused by rejuvenation and difference in rock resistances to erosion may be the probable explanations.

Discussion and Conclusion :

The study of the channel slope of the Basin leads to the following discussion and conclusion :

(i) The thalwegs of the streams consist of a series of connected segment, "each differing those that adjoin it, but all closely related parts of one system" (Mackin, 1948 : 491). It is further stated that each segment has the slope that will provide the velocity required for transportation of all of the load supplied to it from above, and this slope is maintained without change as long as controlling conditions remain the same.

(ii) The thalwegs of the most of the rivers show 'up - concavity' (Fig.14), Gilbert (1877) explained it as the effect of increasing stream discharge. His laws of declivity (gradient) bears an inverse relation to discharge because, as discharge increases, channel cross section increases reducing proportionately the frictional losses of the stream and enabling it to carry its bed load on a lesser slope. Yatsu (1955) thinks upconcavity due to decreasing caliber (diameter) of bed load particles down stream. It is suggested that a lesser gradient suffices for the transport of finer bed materials.

These are only partial exlanations. If increase of load and discharge is the cause, it should continue increasing slope. Ahmad (1976 : 52) gave a detailed and lucid explanation for the develop-

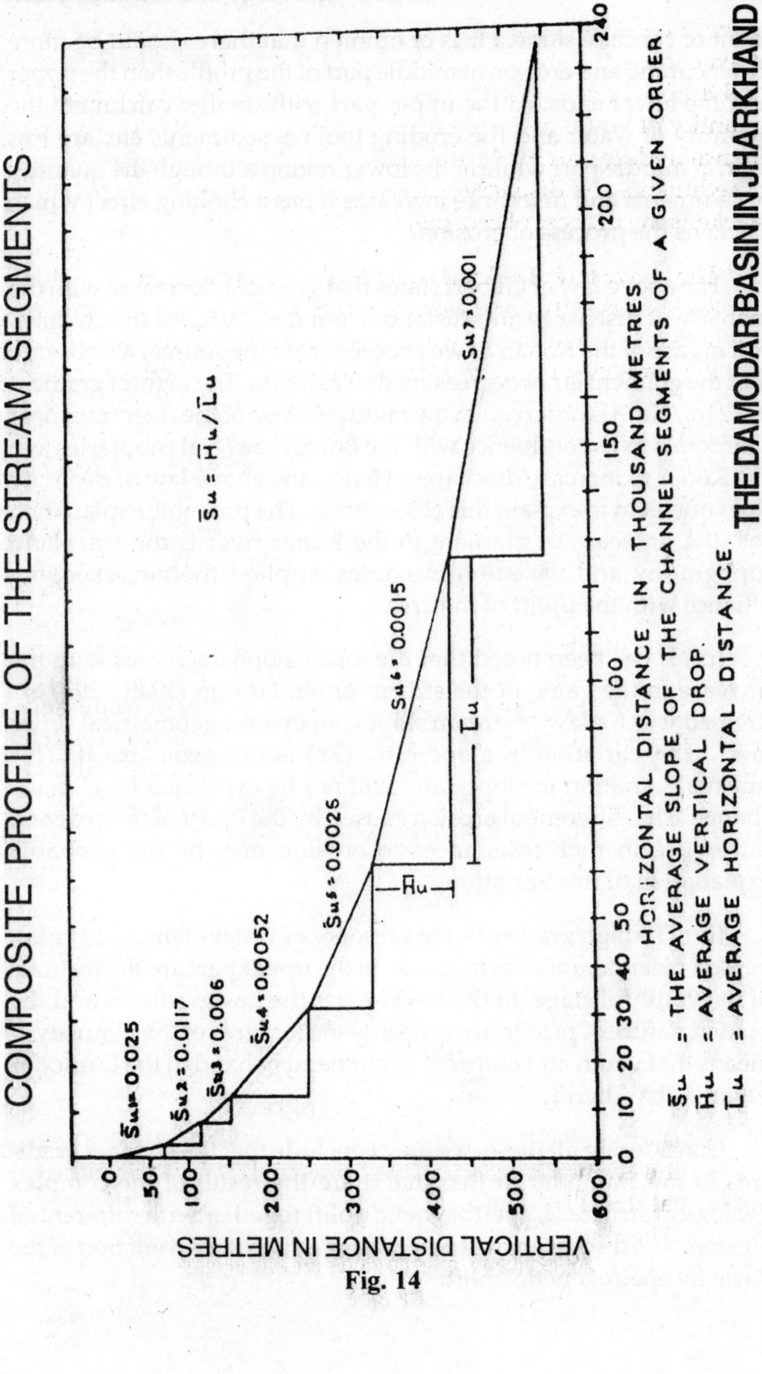
COMPOSITE PROFILE OF THE STREAM SEGMENTS
THE DAMODAR BASIN IN JHARKHAND
$\bar{S}u = \bar{H}u/\bar{L}u$
$\bar{S}u_1 = 0.025$
$\bar{S}u_2 = 0.0117$
$\bar{S}u_3 = 0.006$
$\bar{S}u_4 = 0.0052$
$\bar{S}u_5 = 0.0026$
$\bar{S}u_6 = 0.0015$
$\bar{S}u_7 = 0.001$
$\bar{H}u$
$\bar{L}u$
0 10 20 30 40 50 100 150 200 240
HORIZONTAL DISTANCE IN THOUSAND METRES
0 50 100 200 300 400 500 600
VERTICAL DISTANCE IN METRES
$\bar{S}u$ = THE AVERAGE SLOPE OF THE CHANNEL SEGMENTS OF A GIVEN ORDER.
$\bar{H}u$ = AVERAGE VERTICAL DROP.
$\bar{L}u$ = AVERAGE HORIZONTAL DISTANCE.

Fig. 14

ment of concave slope. He is of opinion that there should be more downcutting and erosion at middle part of the profile than the upper and the lower ends. At the upper part with smaller catchment the quantity of water and the eroding tool i.e. sediments etc. are less than at middle part while at the lower region although the quantity of sediments and discharge increases it has a choking effect which slackens the process of erosion.

The above law of Gilbert states that gradient decreases with the increase of discharge and distance from the source of the streams. But in case of the Konar, as we proceed from the source, we observe that the gradient is not decreasing downstream. The highest gradient (6.72 m./Km.) is observed in the middle course of the river from near Mahtotola to its confluence with the Bokaro. Several tributaries join the Konar to increase discharge. Hence, the above law of declivity does not seem to explain this observation. The probable explanation for the increase of gradient in the Konar river is the scarpland topography and variation in energy applied therein - keeping alliance with the uplift of the area.

(iii) It has been noted that the mean slope decreases with the increase in the value of the stream order. Horton (1945 : 295) expressed it as " a law of stream slopes, an inverse geometrical series law". The variation in slope ratio (Rs) is not geometrical. This unusual variation in slope ratio will not be explained by climatic change. The differential erosion caused by the uplift of the area and difference in rock resistances to erosion may be the probable explanation of this variation.

(iv) The high gradients, the presence of water - falls, pot - holes, incised meanders and terraces etc. in the upper part are the features of the youthful stage. In the lower parts the low gradients and the graded nature of profile are probably the features of the maturity. It means that youth and maturity occur hand in hand in the Damodar basin in Jharkhand.

Considering all these, we may conclude that the channel gradients in the Damodar in Jharkhand are the result of the complex geological structure, epeirogenetic uplift together with differential erosion. The discussion is strongly suggestive of rejuvenation of the Basin by epeirogenetic uplift.

7.1.2. Cross Profiles :

The cross - profiles of the valley shed considerable light upon the geomorphological history of the area. They also indicate the influence of local geological structure over the evolution of the forms of the profiles.

The cross - profiles reveal the following facts :

(i) The cross - profiles of the Damodar and the Barakar valleys are characterized by narrow steep sided gorges, 'V' - shaped valleys and roughly 'U' - shaped valleys. These changes in the forms of the profiles of the valleys are really a function of the nature of rocks and the age of the valleys.

(ii) The cross - profiles of the valleys show ásymmetric slopes. This nature of the valley side slopes may be explained due to variation in energy - keeping alliance with the uplift of the area and also due to differences in geological structure and rock resistances to erosion. Meandering may also be responsible for asymmetric valley side slopes. The concave bank generally has steep slope while the convex side has relatively gentle slope.

(iii) The examination of cross - profiles reveals several 'knick - points'; The profile curves are 'composite' which show the evidence of the rejuvenation of the area. The valley - in - valley at Rajrappa is an excellent example of the same. We have another notable example at Bermo where the Konar dam has been constructed.

(iv) Narrow steep - sided gorges are observed in the area. They are generally seen infront of waterfalls. The Bokaro gorge on the Bokaro river, the Chandro gorge on the Konar river and the gorge on the Dhangarha river in front of Kanti waterfall are some of the notable examples.

(v) The genesis of gorges may be (i) due to recession of waterfalls and (ii) the high rate of vertical corrosion of the river after rejuvenation of the area. Hence, the presence of gorges suggests that the area has been uplifted in recent past.

(vi) One very important point regarding the development of the gorges, is joint in the rocks. Gorges are generally seen along joints. For example, an impressive gorge at Chandro on the konar river is seen developed along joints in the rocks. Similar example may be observed near the Usari waterfall on the Usari river about 15 Kms. south from Giridih. It is important to note that at both these places waterfalls may be observed across joints in rocks.

(vii) In the lower section of the valley, the river beds are generally flat with almost steep banks. This section is characterized by almost 'U' - shaped valleys.

(viii) It is also important to note that some residual hills on two sides of the valley owe their preservation to their being more resistance to erosion. This also provides relatively steep valley - side slope.

(ix) The above discussions conclude that the cross - profiles of the valley are characterized by narrow steep sided trenches. 'U' - shaped valleys, terraces, valley - in - valley, and asymmetric valley - side slopes.

These forms strongly suggest that they are due to the variation in energy - keeping alliance with the rejuvenation of the area.

7.1.3. Surface Gradients :

Surface gradients, "the fundamental unit of the physiographic landscape" (Twidale, 1973 :93) is intimately related to our daily life. It controls the traffic movements, agricultural developments, settlements, industrial developments and military operations. Hence, it is essential to study the slope, its nature and characters, evolution and forms for both its academic value and practical aspects. The problem is complex. To show the slope conditions over the entire Basin of the Damodar in Jharkhand an Average Slope Map (Fig. 15) has been prepared following Wentworth's method* (1930). The map shows the distribution of average inclination of the surface. Categorywise area has been measured and frequency and cumulative frequency have been calculated (Table : 7-II) and plotted (Figs. 15A & 15B). This technique, though, for the construction of slope map is extremely time - consuming, seems to be very useful in making quantitative generalizations of the surface gradients.

Figs 15, 15A, 15B and Table : 7-II reveal the following facts :

(i) The average slope of the Basin ranges from less than 1^0 to more than 14^0. The highest slope is observed in the Parasnath area ($14^0 15'$), followed by Lugu Pahar, ($11^0 45'$) and Baragaon Pahar ($11^0 35'$).

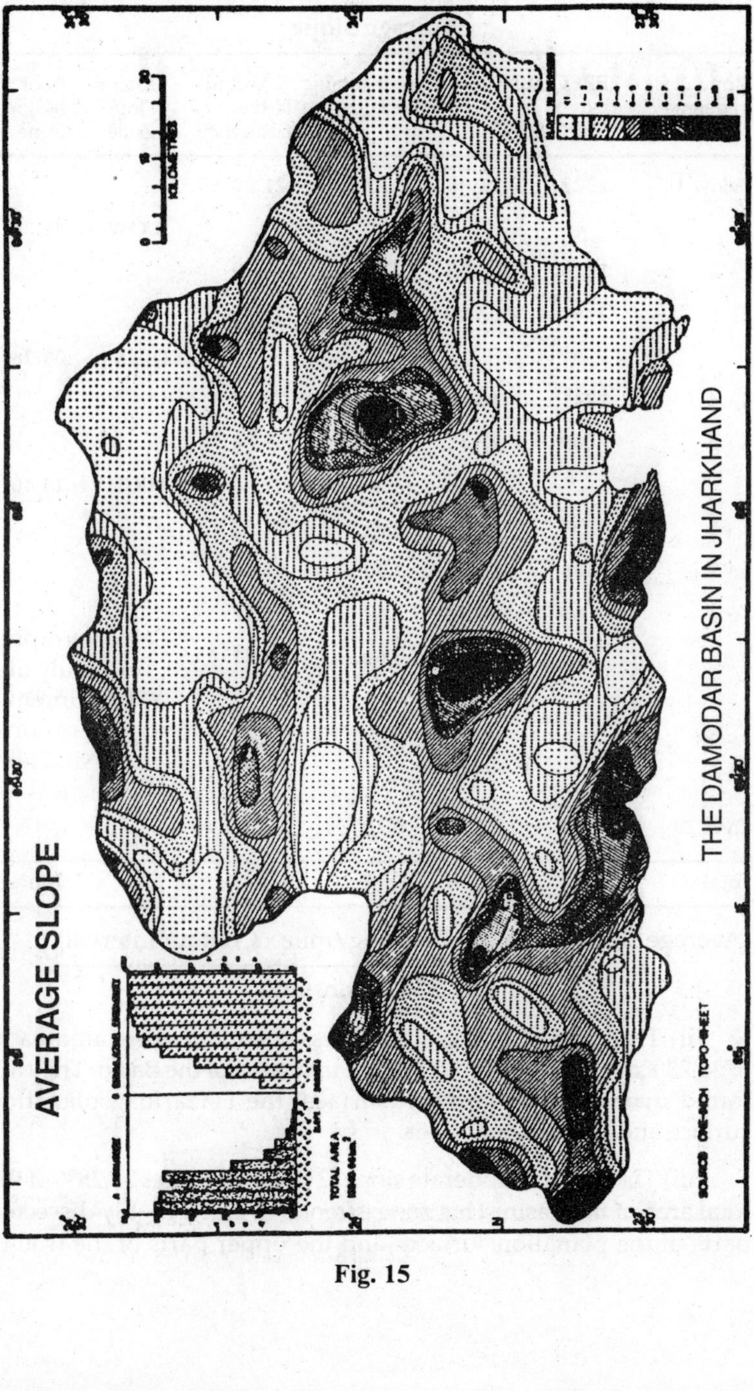
AVERAGE SLOPE
THE DAMODAR BASIN IN JHARKHAND
SOURCE ONE-INCH TOPO-SHEET

Fig. 15

TABLE : 7 - II

Average Slope :

Slope class in degree	FREQUENCY Mile2	Km.2	% of total area	% Cumu lative frequency	Major slope Zone	% of Major Zone
Below 1	1322	3423.98	21.07	21.07		
					LOW	41.39
1 - 2	1275	3302.25	20.32	41.39		
2 - 3	1161	3007.00	18.50	59.89		
					MEDIUM	38.28
3 - 4	1241	3214.18	19.78	79.67		
4 - 5	609	1577.31	9.71	89.38		
					HIGH	14.40
5 - 6	294	761.46	4.69	94.07		
6 - 7	188	486.92	3.00	97.07		
7 - 8	88	227.92	1.40	98.47		
8 - 9	40	103.60	0.64	99.11		
					VERY	5.93
9 - 10	32	82.88	0.51	99.62		
					HIGH	
Over 10	24	62.16	0.38	100 %		
Total :	6274	16249.66	100 %			100%

$$\text{(Average Slope} = \frac{\text{Contour Crossing / mile x Contour Interval)}}{\text{3361 (Constant)}}$$

(ii) The zone of low gradient (less than 2^0) covers an area of 6726.23 Km.^{2} which is 41.39 % of the total area of the Basin. They are found over Ranchi planation surface, the Hazaribag planation surface and the Dhanbad area.

(iii) The zone of Moderate slope (2^0 to 4^0) occupies 38.28% of the total area of the Basin. This zone extends over the highly dissected parts of the planation surfaces and the upper parts of the trough area.

(iv) The zones of high (4^0 to 6^0) and very high (over 6^0) gradients cover 14.4% and 5.93% of the total area of the Basin respectively. They extend over the scarp zones and hilly areas of the Basin.

(v) The general distribution of slope over the entire Basin reveals the fact that the areas of high gradients are generally associated with scarp zones, hills and ridges. Deep - cut river valleys are also responsible for high gradients. The nature and character of rocks, their resistances to erosion, etc. add to the variation of the distribution of slope angles. For example the easily erodable sedimentaries, e.g. shales form usually low lands and also relatively low gradients. Sandstones form bold topography (e.g. Morcha Pahar and Lugu Pahar) and naturally high gradients.

(vi) The areas of planation surfaces due to their flatness owe low surface gradients, though they are of granites and gneisses. The zone of low gradients generally coincides with the zone of low relative relief (Fig. 17).

The above discussions conclude that the vast area (79.68%) of the Basin is under low to moderate slope and only 20.33% of the total area is under high and very high slope. This distribution of slope, probably, suggests that the area has been uplifted in the recent past epeirogenetically and the uplifted planation surface which bear testimony of the earlier erosion surface have practically low gradients and low relative relief even at the higher elevation.

7.1.4. Morphological Character of Hill Slopes :

For morphological analysis of the hill - side slopes in the Basin, sample study of some of the hills has been necessitated. Direct measurement has been conducted in the field selecting ten sites which are significant and accessible. They are mostly hills and hillocks namely Bheldhari, Murli, Bangla, Christian, Poraiya, Chandri, Doma, Lakmi, Koya and Kurmi. These hills are situated in the northern part of the Damodar Basin in Jharkhand. Slope angles have been measured at break points with Abney Level. Distances between ranging rod locations have been measured with tape. Data (Table : 7-III) have been plotted . The points have been joined by straight lines (Fig. 16).

Following the method, suggested by Young (1963, 1964), slope

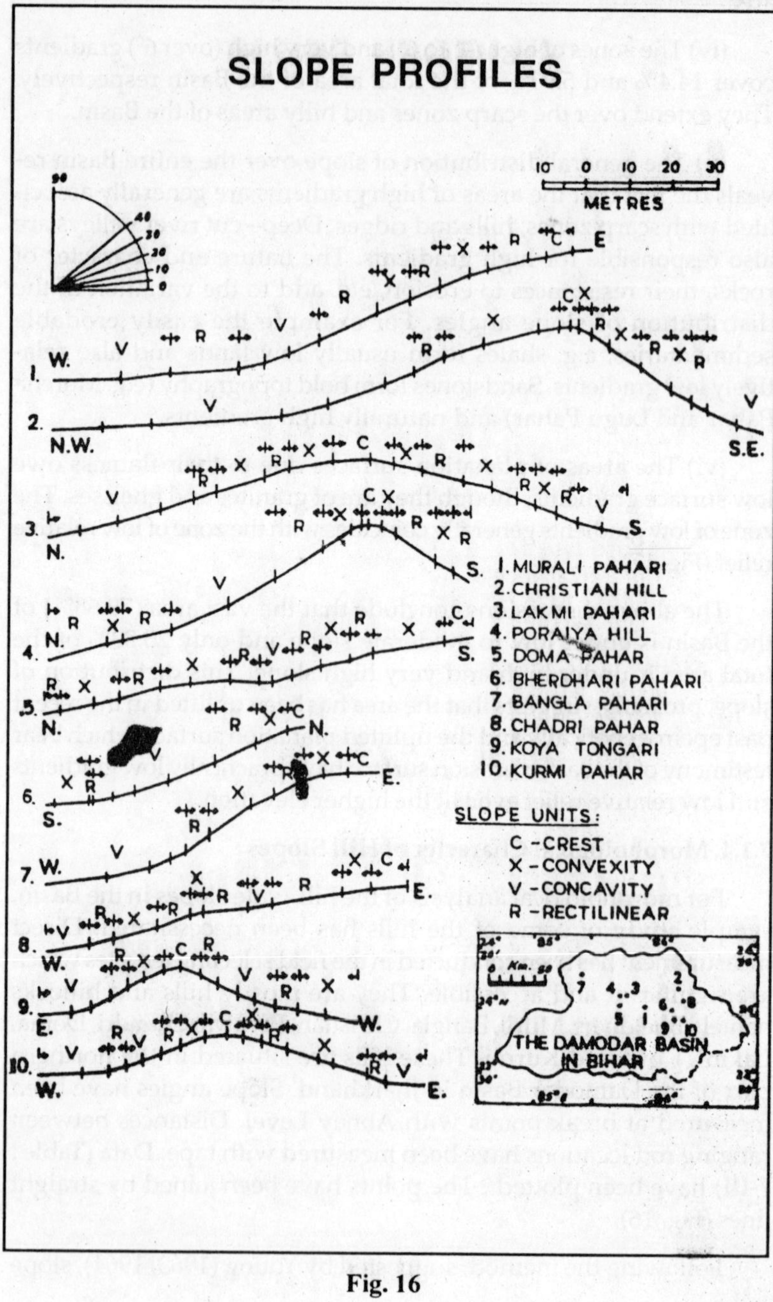
SLOPE PROFILES
10 0 10 20 30
METRES
1. MURALI PAHARI
2. CHRISTIAN HILL
3. LAKMI PAHARI
4. PORAIYA HILL
5. DOMA PAHAR
6. BHERDHARI PAHARI
7. BANGLA PAHARI
8. CHADRI PAHAR
9. KOYA TONGARI
10. KURMI PAHAR
SLOPE UNITS:
C - CREST
X - CONVEXITY
V - CONCAVITY
R - RECTILINEAR
THE DAMODAR BASIN IN BIHAR

Fig. 16

units have been distinguished into crest, concave, convex and recti-linear.

TABLE :7 - III
Field Measurement of Slope Profiles *

CHRISTIAN HILLS			MURLI PAHARI		
Distance in Metres	Angles Degree	Angles Minutes	Distance in Metres	Angles Degree	Angles Minutes
From Foot to Crest			**From Foot to Crest**		
19.4	2	30	26.8	1	50
19.3.	3	00	19.3	6	20
51.2.	3	20	27.7	3	10
18.2	15	30	9.	12	30
23.6	23	10	9.9	20	30
12.6	24	10	12.5	24	50
12.0	29	10	9.5	25	20
16.0	18	30	14	23	50
22.2	16	30	12	23	50
8.3	30	30	13.1	24	10
4.1	3	20	6.9	20	10
			16.5	15	30
			7	6	10
			11.5	9	30
			8	3	10
From Crest to Foot					
3.1	1	30			
5.3	24	30			
8.5	34	00			
15	33	00			
15	37	30			
8	26	00			
7	33	30			
13	24	30			
11	16	40			
31	5	30			
BANGLA PAHARI			**BHELDHARI PAHARI**		
From Foot to Crest			**From Foot to Crest**		
25	8	10	3.2	11	20
17	12	40	13.9	2	10

8.5	23	40	1	40	30
6.9	25	30	8.3	6	40
11	37	40	9	15	20
20	35	30	14.1	20	20
10	24	30	20.1	23	10
12	25	10	17.8	25	40
8	16	40	14	25	10
9.4	7	30	9	9	10
12.7	2	30			

PORAIYA HILL

Distance in Metres.	Angles Degree	Minutes
From Foot to Crest		
24.1	17	10
6.7	13	30
19.5	11	30
24.3	22	10
21.4	30	00
22.4	33	30
11.7	32	30
4.3	23	30
2.7	12	30
From Crest to Foot		
5.4	10	30
6.4	28	00
18.0	35	00
7.4	30	50
7.9	37	30

CHADRI PAHAR

Distance in Metres	Angles Degree	Minutes
From Foot to Crest		
17.5	10	50
10.3	15	40
17.7	14	50
13.2	21	50
5.9	13	40
16.6	5	10
10.1	9	40
27.5	10	10
11.0	8	30
14.0	4	40

LAKMI PAHARI

From Foot to Crest		
12.8	10	30
15.8	14	00
17.4	16	30
14.0	18	00
9.7	19	00
19.2	17	00
10.0	20	00
16.4	10	00
14.2	4	00
From Crest to Foot		
9.5	12	30
11.6	12	00

DOMA PAHAR

From Foot to Crest		
6.0	14	10
13.7	12	30

10.1	12	30	16.2	14	30
16.6	17	10	11.0	13	00
12.5	8	30	10.7	20	00
15.8	8	40	7.3	22	30
11.3	4	30	8.8	20	00
5.3	4	50	9.2	22	00
8.1	16	20	9.2	14	00
12.0	21	10	10.3	10	00
10.9	10	50			
9.7	19	00			
27.8	9	50			
11.3	8	40			

KOYA TONGARI			KURMI PAHAR		
Distance in Metres.	Angles Degree	Minutes	Distance in Metres.	Angles Degree	Minutes
From Foot to Crest			**From Foot to Crest**		
13.1	6	00	16.8	3	30
8.3	17	00	10.3	5	00
9.2	26	00	7.9	19	00
7.9	32	30	7.0	17	00
7.3	28	00	7.3	21	00
8.5	22	00	7.9	23	00
7.0	13	00	10.7	22	00
			8.5	7	00
From Crest to Foot			**From Crest to Foot**		
8.3	20	00	6.7	6	00
7.9	26	00	7.3	12	00
8.3	22	00	10.7	16	00
7.3	21	30	12.5	15	00
7.9	15	00	6.1	15	30
8.8	7	00	13.4	15	30
11.0	11	00	7.3	22	00
11.9	13	00	15.8	2	00

Morphological Character:

The study reveals the following morphological characters of the hills :

(i) Most of the hills in the Basin are asymmetrical in nature, meaning thereby that the slope of one side differs considerably from the other .

(ii) Steep slopes are considerably more extensive the gentle slopes.

(iii) Each slope profile is characterised by convexity, concavity and rectilinear units, thus it is of composite nature and may be called 'composite profile' (Prasad, 1979, 289). It is a general property of the profiles. However, when they are observed minutely the units do not seem to follow this general sequence, rather they have been 'non - sequential'.

(iv) It was observed in most cases that the upper profile was convex, but the lower slope was clearly concave with rock fragments and at places big boulders also. At the foot of the hills, talus materials were observed. At the foot of the surrounding areas a mantle of red soil covers the slope.

Discussion and Conclusion :

The measured profiles are not equally uniform in their slope variation from foot to crest. Convex and concave segments both include those units which are rectilinear sections or maximum segments of profiles (Fig. 16). The degree of curvature of convex and concave slope segments has been influenced by the presence of maximum units, their length and amount of angularity. Its presence or absence determines the complexity of profiles and the nature of development.

These characteristics create a problem for the models of landscape development. It is now to be examined whether the composite nature of slopes of the hills of the Basin is due to slope decline (Davis, 1909 : 249 - 278), or due to parallel retreat (Bryan, 1940 : Fair, 1947 and King, 1957) or related to any other model.

The model of Davis (1909 :249 - 278) interprets that in the sub - tropical climate zone a region is subject to slope reduction generally characterised by convex upper slope and concave foot slope. The measured hill - side slopes are of composite nature.

The Penck - King model envisages the four elements of slope

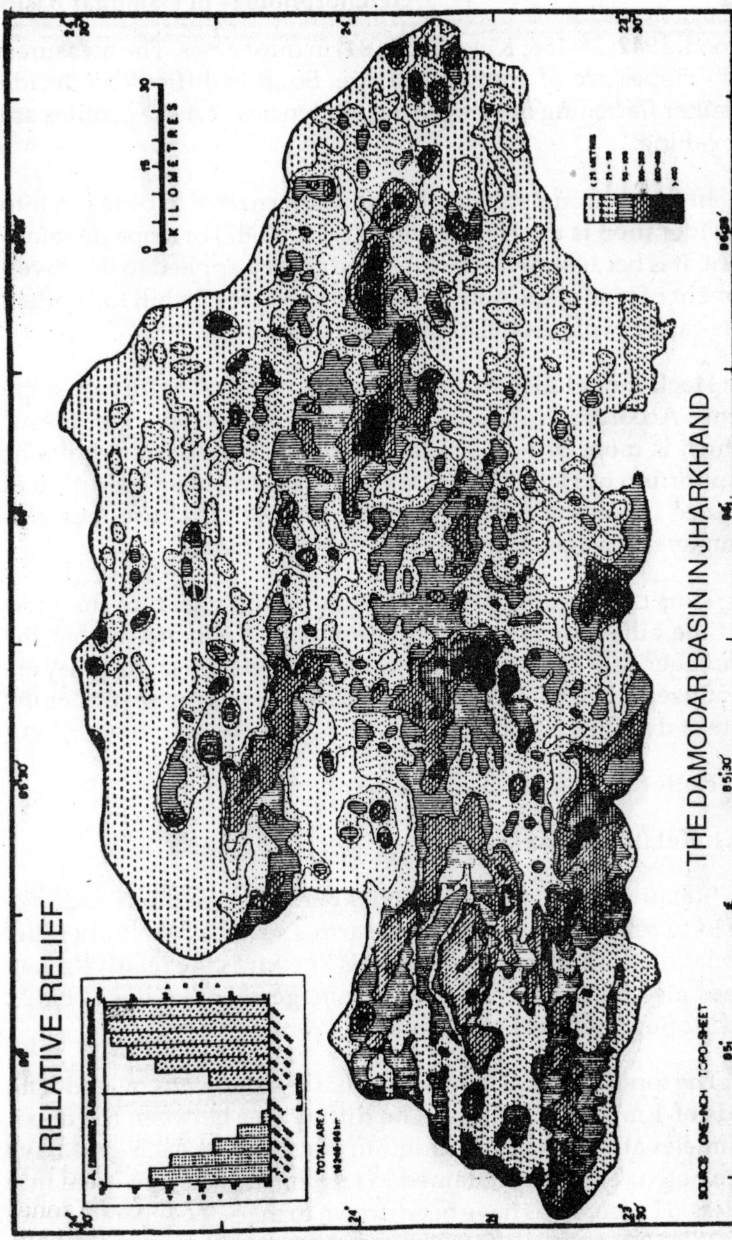
RELATIVE RELIEF
THE DAMODAR BASIN IN JHARKHAND
KILOMETRES
TOTAL AREA
SOURCE: ONE-INCH TOPO-SHEET

Fig. 17

(Wood, 1942: 28-138; King, 1957 : 84) in most cases. The measured hill - slopes are of complex nature. So, it is difficult to decide whether flattening or steepening tendencies of these profiles are prevailing.

In these circumstances, the only alternative to be taken into consideration is the Hack model (1960, 80-87) of slope development. It is because of the fact that the energy applied to the development of landscape may vary from one side of a hill to another. This easily explains the asymmetry of the hills.

Hack model also explains the 'composite' nature of slope forms. According to Strahler's 'equilibrium - slope hypothesis' (which is more or less similar to the slope system in dynamic equilibrium of Hack), the slope angle represents a condition of balance involving not only rock type and structure, but also climate, soil, vegetation and other circumstances.

Thus concluding our discussion, we may lean to the view that the hill - side slopes of the Damodar Basin are neither the consequence of slope - reduction. nor of slope retreat, but they are greatly related to the Hack model of slope development under the state of dynamic equilibrium.

7.2. Relief Measures :

7.2.1. Relative Relief :

'Relative relief' (Smith, 1935) is one of the methods to depict the local relief of any part of the earth's surface. We study relief in relation to adjoining areas, so it is known as the relative relief. It has also been expressed as "drainage relief" (Glock, 1932), and "topographic relief" (Johnson, 1933).

The topographical maps of the Basin have been covered with grids of 1 mile^2 (2.59 Km.^{2}). The differences between the maximum elevation and the minimum elevation in each grid have been noted. The data, obtained in feet, have been converted into meters. The isopleths have been drawn to mark the separate zones (Fig. 17). Categorywise frequency and cumulative frequency have been calculated (Table : 7-IV) and plotted (Figs. 17A & 17B).

TABLE : 7 - IV
RELATIVE RELIEF*

Categories	Relative Relief in Metres.	Frequency In Sq Mile	In Sq. Km.	% of the total area	% of cumulative frequency.	Major categories	% of Major categories
Ist.	25	2307	5975.13	36.83	36.83		
						LOW	69.26
2nd.	26-50	2031	5260.29	32.43	69.26		
3rd.	51 - 100	1102	2854.18	17.46	86.72		
4th.	101 - 150	460	1191.40	7.33	94.05		
						MEDIUM	27.67
5th.	151 - 200	181	468.79	2.88	96.93		
6th.	201 - 250	82	212.38	1.31	98.24		
7th	251 - 300	36	93.24	0.57	98.81		
						HIGH	2.58
8th.	301 - 350	20	51.80	0.32	99.13		
9th	351 - 400	24	62.16	0.38	99.51		
10th.	401 - 450	12	31.88	0.19	99.70		
11th.	451 - 500	1	2.59				
12th.	501 - 550	8	20.72				
13th.	551 - 600	2	5.18				0.49
14th.	601 - 650	3	7.77	0.30	100.00	VERY HIGH	
0.49							
15th.	651 - 700	3	7.77				
16th.	Above 700	2	5.18				
Total :-	6274		16249.66		100.00		100.00

* Relative Relief = Max. Elevation - Minium Elevation.

Figs. and table : 7 - I reveal the following facts :

(i) Fig. 17 clearly shows the differences in relief in the Basin from place to place, without reference to total elevation of the area. The latter (elevation) is relatively "insignificant for the roughness of the area, for hilly country or level plains can occur at many different heights" (Dickinson, 1969). We observe low relief even at the high planation surfaces and high relief in the area of highly dissected terrain relatively of lower elevations.

(ii) The relief ranges from less than 25 metres in a grid of 1 mile^2 (2.59 Km.^{2}) to more than 700 metres in a grid of 1 mile^2 (2.59 Km.^{2}). The minimum relief is 7 metres and the maximum relief is 778 metres in a grid of 1 mile^2 (2.59 Km^2.). But these extremes are limited in a few grids.

(iii) The number of grids with very high relief lies in the areas of high hills. The Baragaon Pahar in the southern scarp zone, Parasnath hill area, Lugu Pahar, Jhumara and Khamarpat areas are the regions of very high relative relief.

(iv) The area having relative relief more than 400 metres in a grid of 1 mile^2 (2.59 Km.^{2}) occupies only 0.49 % of the total area of the Basin.

(v) The area of high relative relief 201 to 400 metres in a grid of 1 mile^2 (2.59 Km.^{2}) occupies 2.58 % of the total area of the Basin. This zone corresponds the areas of scarps and relatively low hills.

(vi) The zone of moderate relief is seen in the areas adjoining the isolated hills and ridges. Relatively gentle scarp zones have also moderate relief. It occupies 27.67 % of the total area of the Basin.

(vii) A vast area (69.26 % of the total area) of the basin has relief less than 50 meters in a grid of 1 mile ^{2} (2.59 .km.^{2}) . They are associated mostly with the planation surfaces of Ranchi , Hazaribag and broken country of the Dhanbad District .

(VIII) High and very high relief are observed in the areas of scarps, isolated hills and ridges, probably not the features of old age of the first denudation cycle, rather they have been rejuvenated or they are intrusive in character.

(ix) The planation surfaces are upland plains and valleys are narrow and steep - sided. But relief is not high. This indicates that the rivers are being incised after rejuvenation of the area.

The relative relief of the Basin has close relation with the daily life of human beings. The areas of high and very high relief are negative zones from the economic and cultural point of view. The area of low relief is suitable for agriculture, construction of roads and railways, irrigation canals, settlements and industrial developments etc.

Here the most important point is to examine the process of evolution of the landscape which sometimes converts even favorable grounds into bad land topography and makes it unsuitable for economic, cultural and industrial developments. For example the process of gully erosion has converted 8.00 % of the total area of the Basin unsuitable for agriculture and other purposes. Thick loamy soil, gentle slope, low relief and monsoon climate are favourable for agriculture. But the evolution and expansion of gullies create problems and the affected areas have been left fallow.

7.2.2. Average Relief :

Average Relief is " the expression of intensity of relief of a particular area" (Prasad, 1979 : 261). It depicts the average or mean value of the sum of the maximum and the minimum elevations with reference to the sea level. Therefore, it gives a somewhat different picture of the landscape of the area.

The objective of this study is to analyse the average relief of the area. With the above intent and purpose, data have been extracted from One - Inch Topographical Sheets (Table : 7 - V) to prepare map (Fig. 18) and histograms (Figs. 18 A and 18B).

Table and Figs. reveal the following facts :

(i) The average relief of the Damodar Basin in Jharkhand ranges from less than 200 metres in a grid of 1 mile^2 (2.59 Km.^{2}) to more than 1000 metres in a grid of 1 mile^2 (2.59 Km.^{2}).

(ii) The zone of low average relief covers 30.83 % of the total area of the Basin. It extends over the eastern part of the Basin where the absolute height and relative relief are also low.

(iii) The zone of medium average relief (301 to 600 m.) occupies the major portion of the Basin. It is 62.47 % of the total area.

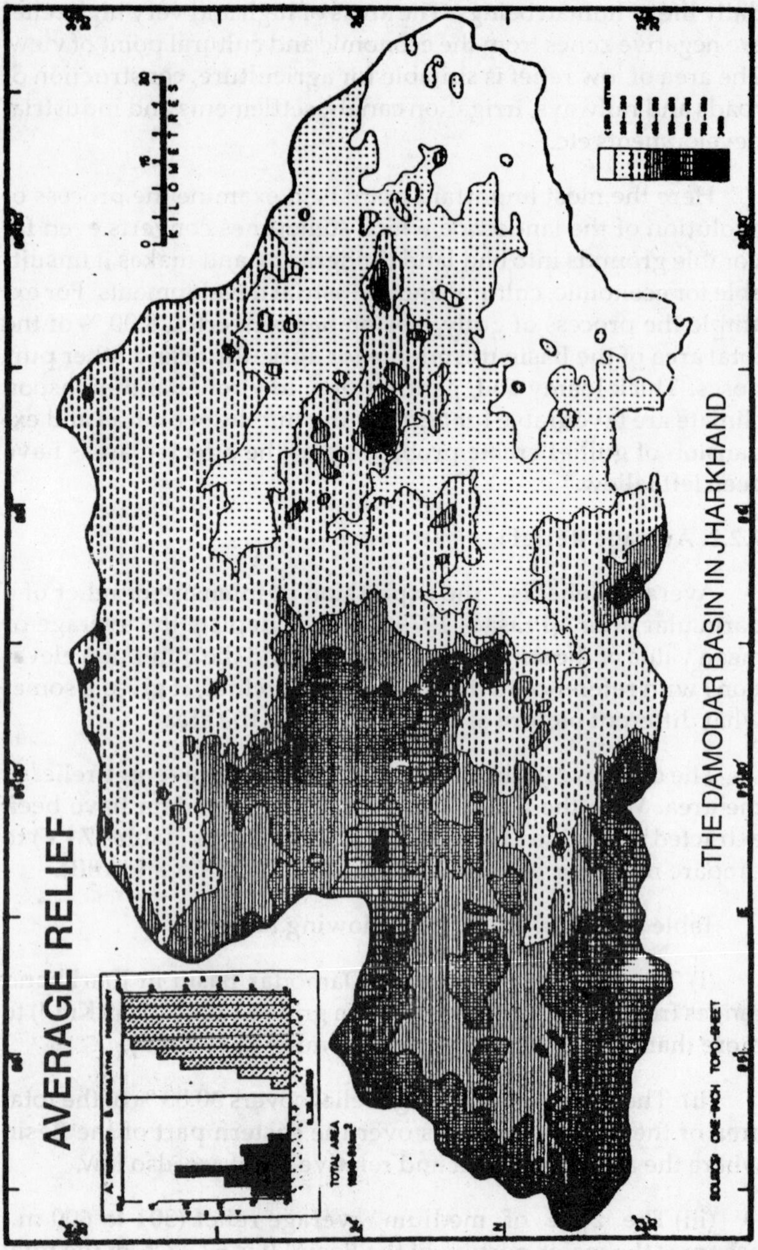
AVERAGE RELIEF
THE DAMODAR BASIN IN JHARKHAND
KILOMETRES
SOURCE ONE-INCH TOPO-SHEET

Fig. 18

TABLE : 7 - V
Average Relief*

Catego-ries	A.R.in Metres	FREQUENCY Sq.mile	Sq.Km.	% of	Mazor zones	% Cumulative % of major zones	
1st.	200	718	1858.62	11.44	11.44	LOW	30.83
2nd.	201 - 300	1217	3152.03	19.39	30.83		
3rd	301 - 400	2137	5534.83	34.06	64.89		
4th.	401 - 500	863	2235.17	13.75	78.64	MEDIUM	62.47
5th.	501 - 600	920	2382.8	14.66	93.30		
6th.	601 - 700	381	986.79	6.09	99.39		
7th.	701 - 800	28	72.52	0.45	99.84	HIGH	6.67
8th.	801 - 900	8	20.72	0.13	99.97		
9th.	over - 900	2	5.18	.03	100.00	VERY HIGH	.03
Total : - 6274			16249.66	100.00		100.00	

$$\text{Average Relief} = \frac{\text{Max. Elevation} + \text{Minimum Elevation}}{2}$$

(iv) The high (601 - 900 m.) and very high (over 900 m.) zones of average relief occur only 6.67 % and 0.03 % of the total area of the Basin respectively.

(v) The highest average relief is found in the Parasnath Hill where absolute relief and relative relief are highest. This is due to intrusive nature of the hill.

(vi) The most remarkable point to be noted is that the absolute relief above 900 metres covers 37 grids. While the average relief occupies only 2 grids in the Basin.

(vii) The comparison of the map of average relief with the maps of absolute relief and relative relief perhaps indicates the difference in the intensity of erosion in the various parts of the Basin after rejuvenation.

7.2.3. Absolute Relief :

Absolute Relief means the maximum elevation of the area with reference to the sea level. For comparative and detailed study of the area 'topographical sheets of entire Basin of the Damodar have been covered with grids of 1 mile^2 (2.59 Km.^{2}). The maximum elevation in each grid has been noted. Data, obtained in feet, have been converted in metres. They have been recorded in the centre of each grid. Isopleths have been drawn to mark the separate zones (Fig. 19). Frequency and cumulative frequency of the area have been calculated (Table : 7 - VI) and plotted in Figs. 19 A & 19B.

Fig. and Table reveal the following facts :

(i) The absolute relief of the Damodar Basin in Jharkhand ranges from less than 200 metres in the eastern portion of the Dhanbad District to more than 1300 metres, in the Parasnath hill, in a grid of 1 mile^2 (2.59 Km.^{2}).

(ii) Parasnath hill (1366 m.), Khamarpat (1068 m.), Baragaon Pahar (1050 m.), Lugu pahar (976 m.) and Jhumara Pahar (934 m.) are the areas of very high absolute relief. They occupy only 0.60 % of the total area of the Basin.

(iii) The zone of high absolute relief (601 - 900 m.) occupies 13.27 % of the total area of the Basin. They are seen in the areas of the Ranchi and the Upper Hazaribag Planation surfaces. Some isolated hills and ridges such as Morcha Pahar, Mahabar hill and Tundi hill areas, etc. have also absolute relief more than 600 metres in a grid of 1 mile^2 (2.59 Km.^{2}).

(iv) The zone of medium absolute relief (301 to 600 m.) covers a vast (61.77 %) area of the Basin. The Giridih - Kodarma plateau, or the lower Hazaribag Planation surface, the western part of the Damodar trough and some small isolated hills, ridges and knolls have elevations between 300 m. and 600 metres above mean sea level.

(v) The zone of low absolute relief, less than 300 metres, occupies 24.36 % of the total area of the Basin. They are seen in the Dhanbad District and the lower parts of the trough area of the Basin.

(vi) The oldest planation surface i.e. the khamar Pat area has the highest elevation and the youngest planation has the minimum absolute heights. They are due to successive and intermittent uplift of the area, during the Himalayan orogeny. But the presence of Parasnath hill, Lugu Pahar, Baragaon Pahar, etc. require separate explanations.

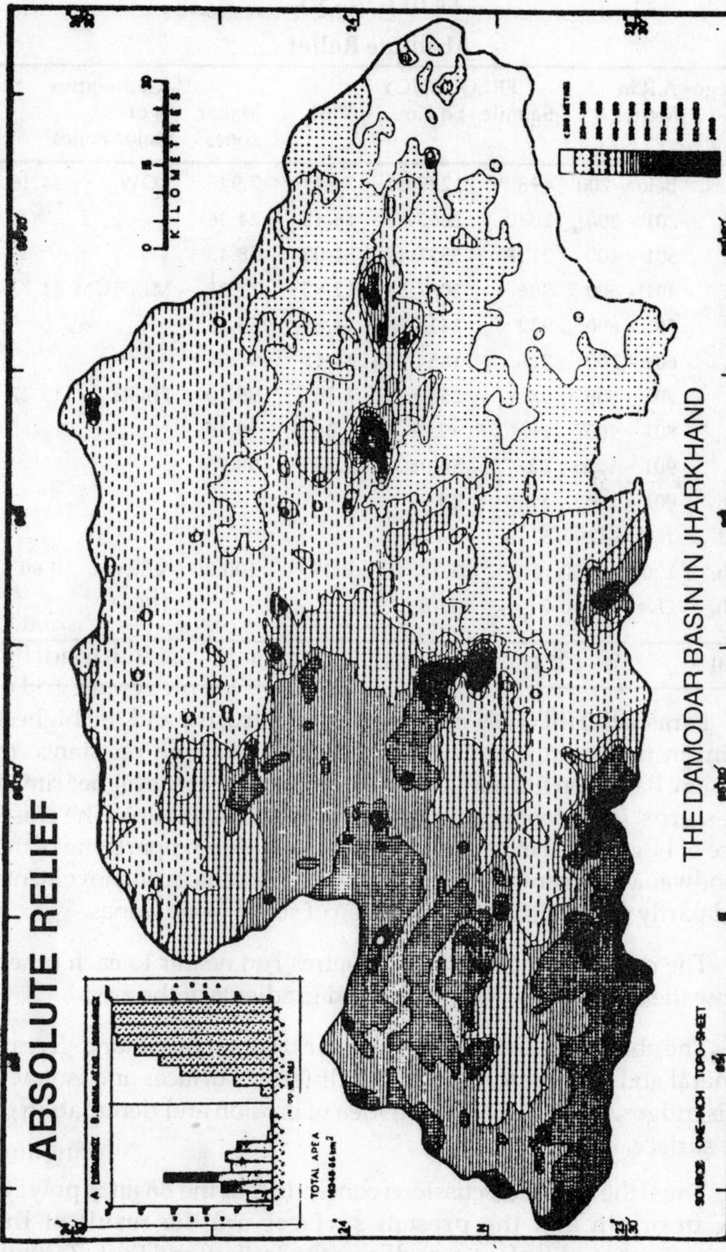
ABSOLUTE RELIEF
THE DAMODAR BASIN IN JHARKHAND
KILOMETRES
SOURCE: ONE-INCH TOPO-SHEET

Fig. 19

TABLE : 7 - VI

Absolute Relief

Catego-ries	A.R.in Metres	FREQUENCY Sq.mile	Sq.Km.	% of	Mazor zones	% Cumulative % of major zones	
1st.	Below 200	498	1289.82	7.94	7.94	LOW	24.36
2nd.	201 - 300	1030	2667.7	16.42	24.36		
3rd.	301 - 400	2138	5537.42	34.03	58.43		
4th.	401 - 500	806	2087.54	12.85	71.28	MEDIUM	61.77
5th.	501 - 600	932	2413.88	14.85	86.13		
6th.	601 - 700	716	1854.44	11.41	97.54		
7th.	701 - 800	98	253.82	1.56	99.10	HIGH	13.27
8th.	801 - 900	19	49.21	0.30	99.40		
9th.	901 - 1000	22	56.98	0.35	99.75		
10th.	001 - 1100	8	20.72				
11th.	1101 - 1200	3	7.77				
12th.	1201 - 1300	2	5.18	0.25	100.00	VERY	0.60
13th.	Over - 1300	2	5.18			HIGH	
Total :-		6274	16249.66	100.00			100.00

Parasnath Hill (1366 m.), an igneous intrusive and the highest point in the Basin, stands like a pillar due to high resistance to erosion. Baragaon hill may partly be recognized as the higher rim of the scarps. It is probably eminences of older mountains. The existence of Lugu Pahar may partly be explained, due to uparching of the Gondwana sediments during the Post - Gondwana earth movements and partly due to differential erosion of softer Gondwanas.

The contours of 500 and 600 metres run nearer to each other, show the characters of the scarps and gradients of the area.

The absolute relief of the Damodar Basin in Jharkhand gives a general and detailed pattern of the different surfaces and isolated hills, ridges, etc. It also gives an idea of erosion and denudation of the surface.

Thus, the above discussions conclude that the Basin is polycyclic in origin and the present surfaces are the result of the epeirogenetic uplift of the area during the Tertiary and Post - Tertiary

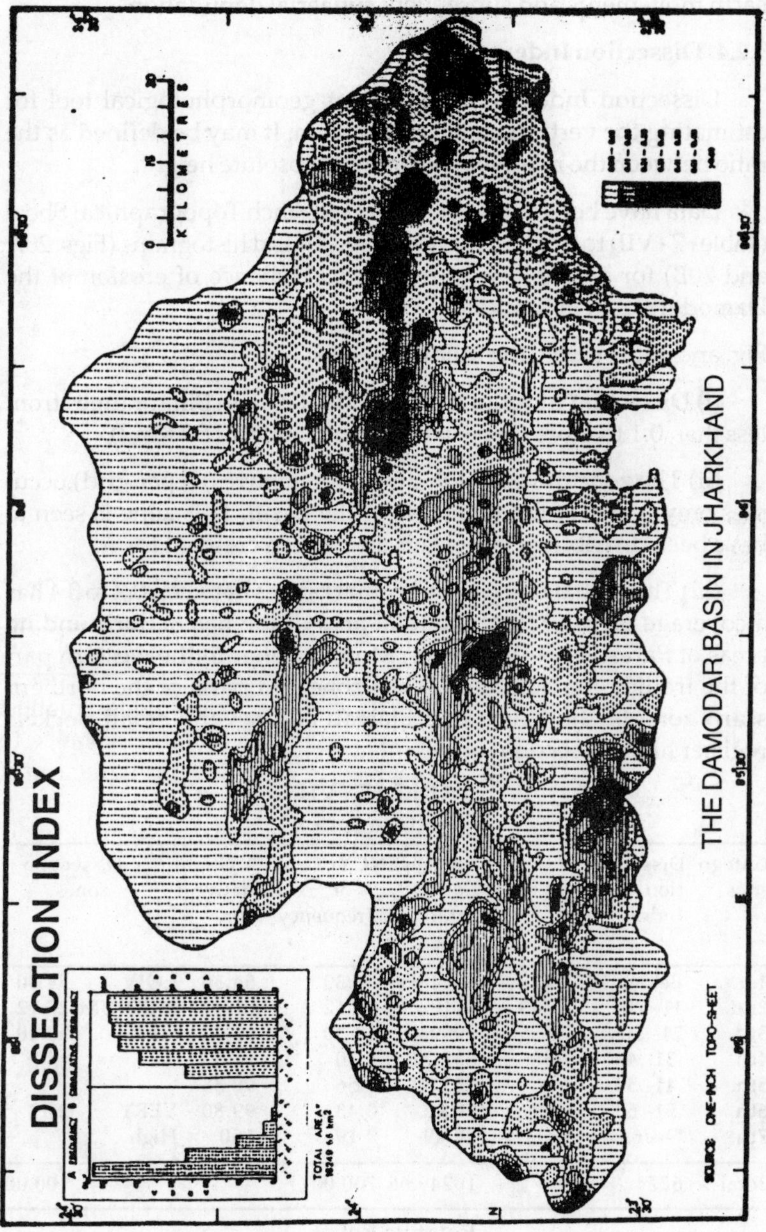
DISSECTION INDEX
THE DAMODAR BASIN IN JHARKHAND
KILOMETRES
SOURCE: ONE-INCH TOPO-SHEET

Fig. 20

earth movements and subsequent subaerial denudation.

7.2.4. Dissection Index :

Dissection Index is an important geomorphological tool for estimating the vertical balance of erosion. It may be defined as the ratio between the relative relief and the absolute height.

Data have been extracted from One - inch Topographical Sheet (Table : 7 - VII) to prepare the map (Fig. 20) and histograms (Figs. 20A and 20B) for an analysis of the vertical balance of erosion of the Damodar Basin in Jharkhand.

Fig. and Table reveal the following facts:

(i) Dissection Index of the Damodar Basin in Bihar ranges from less than 0.1 to more than 0.6 in grid of 1 mile^2 (2.59 Km.^{2}).

(ii) The zone of very high dissection (above 0.4 per grid) occupies only 2.18% of the total area of the Basin. This zone is seen to have been distributed in pockets and also in crescent form.

(iii) The high dissection index that ranges between 0.2 to 0.4 has a coverage of 16.40 % of the total area of the Basin. Surrounding areas of Parasnath - Tundi hills, Morcha Pahar in the western part of the trough, the Southern. scarp zone and parts of the northern scarp zone of the Hazaribag Plateau and several small pockets register high dissection index.

TABLE : 7 - VII
Dissection Index *

Catego -ries	Dissec- tion Index	FREQUENCY Sq.mile	Sq.Km.	% of total area	% Cumu -lative Frequency.	Dissection zone	zones
1st.	Below .10	3415	8844.85	54.30	54.30	LOW	54.30
2nd.	11-.20	1703	4410.77	27.12	81.42	MEDIUM	27.12
3rd.	21-.30	772	1999.48	12.30	93.72	HIGH	16.40
4th.	.31-.40	258	668.22	4.10	97.82		
5th.	.41-.50	98	253.82	1.56	99.38		
6th.	.51-.60	227	69.93	0.43	99.80	VERY	2.18
7th.	Over -.60	11	28.49	0.19	100	High	
Total :-		6274	16249.66	100.00			100.00

$$\text{*Dissection Index} = \frac{\text{Relative Relief}}{\text{Absolute Relif}}$$

(iv) The area of medium dissection index (0.1 to 0.2) covers 27.12% of the total area of the Basin. It may be seen in the belts of relatively gentle scarps and the hilly country of Dhanbad District. Numerous pockets of medium dissection may also be observed in the different parts of the area.

(v) The zone of low dissection index (less than 0.1) covers more than half of the area of the Basin (54.30%). They are seen to have been in the Ranchi Planation surface, the upper Hazaribag and the Lower Hazaribag planation surface, the upper and the middle parts of the Damodar trough. It is important to note that the monotony of low dissection index has been broken by isolated hills and ridges.

(vi) The pattern of isopleths of dissection index is more or less the same as that of the relative relief and absolute relief. But, in the eastern part of the Basin the picture is somewhat different. The area falls under the zone of low absolute relief (less than 300m.) (Fig. 19), but the dissection index is medium (.1 to .2) due to accidented relief.

(vii) The areas of high and very high dissection generally coincide with the belts of high and very high relative relief. They are generally associated with the scarp zones, isolated hills and ridges of the Basin. This, probably, suggests that the area has been uplifted epeirogenetically and the Basin is a multicyclic region.

(viii) The above discussions conclude that most of the Damodar Basin in Bihar is under low and medium (less than .2) dissection index and so the Basin is neither in the old stage nor in the initial stage. It is passing through youth to maturity and we may say that youth and maturity occur hand in hand in the Damodar Basin in Jharkhand.

8

Some Significant Geomorphic Features and their Interpretations

8.1. Scarps :

The scarps are the sharp edges of the plateaus. Different planation surfaces descend to their immediate lower surfaces with steep scarps. They are common landforms in the Basin which attract the attention of sightseers and the research workers in geomorphology.

Scarp, a landscape of exceptional complexity and beauty, overlooks the valley and also the lower planation surface . It covers an area of 2810.05 $Km.^2$ which is 17.28 % of the total area of the Basin. The crown of the scarp is 600 metres in the Ranchi and the Hazaribag plateaus and 1000 metres (AMSL) in the pat region. The hills and hillocks on the scarps add to the general elevation of the scarps and also create complications for their geomorphic interpretations. Numerous river valleys of varying width and depth have broken the edges of the scarps. The rivers generally flow through "dark V - shaped gorges and nearly all the roads which ascend from below follow one or the other river valley and go up by hair - pin bends near these points called 'ghats' (Bose, 1948). The long profiles of numerous streams have been broken by the sudden break in slopes and they (streams) descend the scarps with waterfalls. Kanti waterfall on the Dhangarha river, on the southern scarp of the Damodar trough and Bokaro waterfall on the Bokaro river on the northern rim of the trough may be noted as important examples. The scarps have been dissected into a number of short prominent ridges, the alignment of which is greatly influenced by the tectonic or structural set - up of the area. E.S.E. - W.N.W. trends of the majority of these ridges reflect the predominant Pre - cambrian metamorphics (Chatterji, 1969 : 7). At several places the scarp is a wide zone consisting of a series of scarps rather than a single scarp.

The upper part of the scarp is generally "a vertical palisade of bare rocks while the base is covered up by typical talus slope" (Bose, 1948). The slope varies from less than 5^0 to more than 55^0. The northern scarp of the Ranchi plateau is narrow and steep than the southern scarp of the Hazaribag plateau. The local relief varies from 100 metres to more than 400 metres. It is maximum in the areas of hills and ridges, e.g. the Baragaon Pahar area on the southern scarp zone and the Khamarpat and Birjanga Pahar areas in the south - western corner of the Basin, the Jhumara Pahar area on the northen bank of the Bokaro river and Mahabar hill area in the northen scarp of the Hazaribag plateau, etc.

Plate - 7 : The River through forested scarp

The scarps are generally continuous. But they are provided with gaps and 'ghats' for geomorpphic significance. These gaps and 'ghats' are being used as the important lines for human activities. Tori gap (480 m.) is used for railway and roadway between the Damodar Basin and the Auranga Basin. The national highway from Ranchi to Patna runs through, Chutupalu, Padma and Rajauli (out - side the area) ghats.

Before tracing the genesis and evolution of scarps of the Damodar Basin in Jharkhand the following points may be taken into account :

(i) Most of the area is composed of ancient crystalline rocks which never went beneath the sea.

(ii) During the Permo - Carboniferous earthmovements, a trough was developed in which Gondwana sediments were deposited.

(iii) The entire Damodar Basin with Chotanagpur plateau was reduced near to baselevel before the Himalayan earthmovements took place.

(iv) The whole area was upplifted during the Tertiary and Post - Tertiary earthmovements.

(v) The southern scarp of the Damodar trough is relatively narrower and steeper than the northen scarp.

(vi) The scarps which bound the planation surface are almost circular.

(vii) The arrangement of slopes of the scarp zones agrees to some extent with the four elements of scarps or slopes of Wood (1942) and King (1953) i.e. waxing slope, free face, debris slope and waning slope (Ahmad & Debi, 1965 : 81 - 85, & Debi, 1968 : 81 - 93). But measured profiles of hill slopes show their composite nature (Gupta, 1980).

(viii) The edges of the scarps are broken. Hills and hillocks are seen to be present on the scarps. When we go to the edge of the plateau, we have to go up before descending down the scarp.

(ix) A distant look present a view of the single scarp with dark steep slope. But on approach a series of scarps may be observed.

Genesis :

It is a complex problem to trace the origin and evolution of the scarps. What is more complicated is to discuss the controversial models of the origin of the scarps of different scholars, e.g. 'the fault scarps' of Dunn (1942 : 12), 'the composite scarps' of Verma (1961:127) and' the fault line scarp' of Ahmad (1958, 1965, 1969, 1970).

Dunn, while describing the uplift of the Ranchi and the Hazaribag plateaus, writes that " in the latter part of the Tertiary period a trough developed along the valley of the Damodar, presumably due to faulting" (Dunn, 1942 : 12). Thus, according to him the faulting is Recent and if it is established, the scarps may be `fault - scarps'. The frequency of faults within the coal fields may indicate their recent origin. But these faults are intra - basinal and therefore, they may not be taken responsible for the origin of the scarps on the two sides of the trough.

Verma has advanced the view of the 'composite scarps'. She recognised the upper part as the fault scarp and the lower part as the erosional scarp. It appears to be unacceptable. Because, she writes that "straight boundary was formed by faulting before the deposition of Gondwana beds" (Verma, 1961 : 127). Evidently, she accepts the early formation of the trough and in that case the question of fault scarps on the upper part does not arise.

Ahmad recognised the scarps on either side of the Damodar trough as ' the fault line scarps'. The Gondwana trough was developed during the Permo - Carboniferous earthmovements. The area was reduced to peneplain before Tertiary orogeny took place and according to him there would have been no scarps , etc. on such a surfaces. During the Tertiary and Post - Tertiary periods the entire region was uplifted. The softer Gondwanas were eroded away rapidly and scarps developed due to differential erosion on the two sides of the fault line. Hence, it may be called as ' fault - line scarp ' (Davis , 1913).

The following observations may be noted in favour of fault - line scarps:

(i) There is close correlation between topography and rock resistances on two sides of the fault - line. The softer and less resistant sedimentary rocks of the trough areas have been easily eroded away and resistant rocks form bold relief. If it would have been fault scarp, there would have been no such relation in the evolution of landscape on the two sides of the fault - line, the nature and the character of the rocks would have been more or less the same.

(ii) The southern scarp of the Hazaribag plateau is not as steep and narrow as the northen scarp of the Ranchi plateau. The reason for this is not apparent. But, this probably is very strong pointer to the fact that the Damodar valley is not due to the recent faulting and scarp on either side of the trough is not the fault scarp but fault - line scarp. If the faulting were recent, the steepness of scarps on either side would have been maintained.

Secondly, we do not know the inclination of the original faults, which are difficult to trace. It is likely that the fault on the northern side of the trough was gently inclined.

(iii) The drainage lines across the faults have been superimposed.

(iv) If it is not fault - line scarp, it may be 'fault scarp' or 'composite scarp' as suggested by Dunn and Verma respectively. In either case, we have to suppose that the scarps are due to recent faultings, and in that case the area should suffer frequently from severe earthquakes. But from the records we observe that the region is seismically stable (Sarma, 1975 : 53). This is also evidence that three of the recent, earthquakes - the Assam earthquakes of 1897 and 1950 and Bihar - Nepal earthquake of 1934 did not cause any seismic activity in the trough area. The eastern fringe area, probably, due to side effect, experienced minor tremors. The field geologists have also declared this belt as tectonically stable and faults as "fossil faults".

(v) Describing the scarps of Colorado area, Blackwelder (1934) and Longwell (1946) have indicated that most of the scarps are fault - line scarps and that the faulting was of ancient rather than recent age and thus antedated the period of peneplanation.

Similar is the case with the Damodar trough area in Jharkhand. The faultings took place well before the peneplanation of the region and present scarps are due to the rejuvenation and reexcavation of the area. Thus, we conclude that the scarps on either side of the trough are fault - line scarps. They are the second cycle phenomena and the topography is secondary.

The question of the scarps which occur over different planation surfaces partly differs from the interpretation of the fault - line scarps (discussed earlier). The problem exists whether these scarps are due to local faulting and differential movement, or local warping (Dunn) or 'scarp - retreat' (Penck King and applied to the area by Ahmad) or this is a 'erosion - scarp' of past pediplain (suggested by Pecsi, et al. during the Symposium on Erosion Surfaces in Dec. 1968).

The differential erosion (Chatterji, 1946) does not appear to explain the genesis and evolution of these scarps as the rocks on the scarps and on the planation surfaces are of equal resistance.

The concept of J.A.Dunn, that these scarps are due to successive uplift of the area, seems to be acceptable. But when he regards them to be fault - scarps, the difficulty of visualising the faults along the scarps comes in the way.

Ahmad thinks these scarps are due to parallel retreat. This concept has the following merits :

(i) It does not face any difficulty in proving the faults in the field along the scarps.

(ii) The successive regional uplift of Chotanagpur along the same line i.e. along its outer edge, under isostatic adjustment due to erosion is a more likely sequence of events of the region (Ahmad, 1976 : 47).

(iii) "The persistence of the comparative steepness of the slopes of the scarps, instead of their being converted into gentler gradients show parallel retreat" (Ahmad, 1962 : 29).

The arrangement of the slopes of the scarp zone agrees to some extent with the four elements of scarps or slopes of Wood and King i.e. waxing slope, freeface, debris slope and waning slope. But the measured profiles of some of the hills in the area show their composite nature (Fig. 16). This nature of hill slopes is neither related to slope decline of Davis nor to slope retreat of Penck - King and others in all aspects. It is most probably related with the variation in energy - keeping alliance with the uplift of the area. In other words we can say that it is related with dynamic equilibrium.

8.2. Tors :

"Tor is a residual mass of bed rock produced below the surface level by a phase of profound rock rotting effected by ground water and guided by joint systems, followed by a phase of mechanical stripping of the incoherent products of chemical action" (Linton, 1955 : 476).

Many tors have been observed in the source region of the Nalkari river near Pithouria, along the road from Hazaribag to Bagodar and also from Giridih to Mirzaganj. They are of base rock masses surmounted and surrounded by blocks and boulders.

Plate - 8 : Tors at Silwar

Near Birni several hills have been observed on either side of the Giridih - Birni road. These hills have entirely been converted into weathered blocks of different sizes. Horizontal and columnar joints are clearly observed. Similar feature has been observed around Pithouria, about 15 Kms. north from Ranchi and also along the road from Hazaribag to Bhelwara. At several places hills have been worn down to a 'heap of boulders'. Closely spaced joints in the rocks, perhaps, are responsible for this shattered appearance.

These features may be observed almost at every planation surface, i.e. from the Lower Hazaribag plateau to the Ranchi plateau. The Basin falls under monson climate with alternate dry and wet seasons. But rainfall an important elements of climate, varies considerably. The annual rainfall in the Dhanwar area is about 1100 mm, while in the southern scarp zone it is more than 1400 mm. It means the variation in the distribution of rainfall has nothing to do with the development of such landscapes.

Again, the presence of tors almost in every climatic zone of the world has been reported. This supports the contention that climates cannot be the guiding factors for the development of tors.

Various theories e.g. 'two stage process theory' (Linton, 1955), 'Periglacial theory' (Palmer and Nelson, 1962 :315 - 340) and

'Pedeplanation theory' (King, 1958 : 289 - 91), have been put forward to explain the genesis of tors.

The origin and development of tors in the Damodar Basin in Bihar may be explained as follows :

(i) The tors have, generally developed on the granites which are intrusive in character. They have developed "joints because of tensile stresses caused due to the contraction during cooling (Singh, 1977 - 95). The joints have been further developed due to orogenic disturbances.

The processes of weathering and erosion resulted into exhumation of hard and resistant granitic hills and domes. Here deep basal weathering, as suggested by Linton, cannot be overlooked. The rocks were rotted to a greater depth and the materials were removed by the denudational processes. This removal, rather exhumation, resulted into dialation of granite mass due to pressure release leading to recoil of rocks upward and thus the jointdeseased rocks were exposed to further attack (S.Singh, 1977 : 96).

Epeirogenetic uplift of the area during the Himalayan orogeny, removal of debris by continuous processes of mechanical and chemical weathering and erosion may be the probable explanation for the development of tors in the Damodar Basin in Jharkhand.

8.3. Water Falls :

The Damodar Basin in Jharkhand, consisting of various planation surfaces and scarps, is dotted by a number of waterfalls. These are picnic grounds for the general public. They are of much interest for the students of geomorphology as they shed considerable light on the geomorphic evolution of the area.

The genesis and evolution of waterfalls of the Damodar Basin, Jharkhand be traced back the epeirogenetic uplift of the area during the Himalayan Orogeny. The rivers descend the steep scarps created by much uplifts and form waterfalls. The waterfalls are also formed when a vertical mass of resistant rock extends across the bed of the stream. The differential erosion due to several physical factors are also responsible for the development of waterfall sites in the area.

Thus, according to the mode of their occurrence waterfalls of the area may be kept into the following groups :

i) Scarp falls
ii) Discordant falls
iii) Vertical barrier falls.

8.3.1. Scarp Falls :

The root cause of this type of waterfall is the uplift of the area. Uplift has caused the formation of scarps. Such scarps have become the sites for waterfalls. It has been noted earlier that the different planantion surfaces descend with steep scarps to their immediate lower surfaces. So, this type of waterfalls is the most spectacular and the most common. The Kanti waterfall at the Dhangarha river, on the southern scarp of the Damodar valley, is one of the notable examples. The Bokaro river creates a magnificent fall of about 15 metres at the head of the deep V - shaped gorge it has notches in the crystaliines (Bose, 1948). A sheer fall of 305 metres of a tiny stream from the top of the scarp opposite Baragaon excavates a 'bottomless' plunge hole (R.P.Singh, 1969 : 64). Most of the streams e.g. the Bokaro, the Ghaghara, the Konar, the Usari and the Kewta etc. while step-

Plate - 9 : Entrenched Meander & Gorge at Rajrappa

ping down the plateau surfaces, make waterfalls along the line of scarps. The presence of narrow and deep gorges in front of the waterfalls may, probably, be explained by the recession of water falls and also due to the incision of the river channels after the rejuvenation of the area in the recent past.

8.3.2. Discordant Falls :

This type of waterfall is the result of the differential erosion of the river courses. We have typical example of this type of waterfall at Rajrappa about 13 Kms. north from Gola. It occurs at the confluence of the Bhera with the Damodar . The uplift of the area, probably, facilitates the Damodar to carve its bed deeper, so ' a strange phenomenon is its discordant junction, for it looks like a veritable hanging valley of glaciated region" (Khan, 1954). It falls over a 10 - metre edge into the parent stream.

Plate - 10 : Rajrappa Waterfall

The differential erosion in between the two streams is due to the following causes :

(a) The Damodar is the main river. It has greater volume of water and load than the tributary stream, i.e. the Bhera.So., the former performs greater erosion than latter.

(b) The Damodar is perennial river and the Bhera is seasonal in character. It (the Bhera) gets water only during the rainy season. So, the Damodar continues deeping its bed whereas the Bhera becomes inactive during dry season.

(c) The Damodar roughly flows along the joints near the waterfall and the Bhera flows across the dip of the jointed gneiss and granite (Ahmad, 1969).

(d) The gradient of the Damodar is greater than the gradient of the Bhera. This is against "the law of accordant junction". of Playfair. According to him "side streams normally have steeper gradients than the trunk streams to which they are tributary" (Von Engeln, 1960 : 195). The bed of the Damodar is ungraded here. Numerous pot holes and rapid flow of water may clearly be observed in the Damodar river. Relatively gentle slope of the bed of the Bhera is covered with sands. These clearly indicates the greater erosion of the bed of the Damodar than the Bhera.

(e) It is important to note that the slope is being undercut by the Damodar river, the result of which is to increase the slope angle and to make gravitational stress more effective. Small landslides can be observed on the bank of the channel.

The above discussions conclude that the discordant junction is mainly due to the uplift of the area in the recent past.

8.3.3. Vertical Barrier Falls :

A vertical barrier waterfall is formed when a vertical mass of hard and resistant rock lies across a river bed beyond which the rocks are softer. It is most common in the area where igneous intrusion has taken place. In this region many hard rocks have intruded and lie across the channels forming falls and rapids . A notable example of this type of waterfall may be observed in the Siwani river near Pipcho (a small village). This waterfall is caused by the dyke that traversed through its bed. Rapid flow of water may also be observed here.

8.4. Potholes :

Potholes (Alexander, 1932), features of youthful stage and ungraded nature of streams, are numerous in the Damodar Basin in

Jharkhand. They are more or less circular depressions (Plate :10.1) in the bed rocks of the streams worn out by the gyratory abrasion of pebbles or boulders rotated under the energy of moving water (Holmes, 1975). The common occurrences of potholes appear to be favoured by coarse sandstones and granites. The diameters of potholes vary from a few centimetres to more than 5 metres. The depth ranges from a few centimetres to 1 metre or even more.

Plate - 11 : Pot hole at Chandro

The uplift is the root cause for the development of potholes (Fairbridge, 1968). The epeirogenetic uplift of the Damodar Basin in Jharkhand has accelerated vertical erosion energy of the streams. Water jets or eddies of specially high speed direct the abrasive action of suspended fragments or grindles against chosen spots and thus the holes are scooped or gorged out in the rock at these places (Thirunarainan, 1954). The exposed surfaces of sandstones and granites and gneisses in the Sooknad valley near Giridih and in the Konar Valley at Chandro more than 100 potholes have developed within an area of a few hundred sq. metres. The vertical succession of potholes at Chnadro perhaps suggest the successive uplift of the area. Here fantastic example of potholes at the slopes of about 45^0 may be observed. The recession of waterfall may clearly be seen here.

8.5. Spring :

Spring is a continuous or intermittent natural flow of water from underground. Its site depends upon position of the water - table, the shape of the land surface and types of rocks, etc. In the Damodar Basin in Jharkhand, on the steep slopes (Scarps), the underground water - table is exposed towards the valley side and here water emerges as spring, an important source of water near settlements. They also feed the streams below and the river regime is increased.

Hot Springs :

Numerous hot springs in different parts of the Basin may be observed. They have special thermal and mineral character. Thermal springs are those whose water is at higher temperature of the locality. For instance, the tempereture of the Suraj Kund and the Bangahara hot springs is 88^0 C and 50^0 C. respectively.

Regarding the cause and origin of hot springs our knowledge is insufficient. But, high temperature of these springs indicates that the water rises from great depth and so they are 'ascending springs'. The depth of the Suraj Kund hot spring has been calculated about 4000 metres (N.Prasad, 1973 : 19 & 27).

Most of hot springs of the area occur on the faults. They are genetically related with tectonic activities of the area i.e. faults. "Fault or crack provides a link between the surface and deep - seated region of the crust where the water is both hot (because of increasing pressure with depth or because of contact with some igneous intrusive body) and richer in mineral contents" (Ahmad, 1965: 34). Sulphureous water of the Rajrappa and other springs of the area are used for baths. They have medicinal value for skin deseases.

8.6. Gullies :

Gully Erosion is the most spectacular form of erosion in the region. The Damodar Basin in Jharkhand is confronted with moderate to severe gully erosion. It is a problem for agriculture as it depletes the soil of its fertility. It is also a challenge for water management engineering, as it reduces the life of the reservoires constructed for irrigation, hydel power generation and flood control, etc.

Nature of Gullies :

The author observed some of the gullied areas in the Basin during his field work. He came to the conclusion that the gully erosion in the Damodar Basin in Jharkhand is severe along the banks of the streams and at the foot of hills. Gullies are also present within and outside the forest areas along the river banks.

Most of the gullies are U - shaped and V - shaped. Saucer - shaped head gullies are also observed in the area. At Boothfirm (23°38′45"N. 84°57′30"E. / 460metres) there is a beautiful example of gullying of this shape.

Plate - 12 : A impressive view gully confluence

The depth of the gullies varies from less than 1 metre to several metres. Generally, the depth or downward cutting of gullies is controlled by the bed of the streams. Somtimes, bed rocks also play a major role in controlling the depth of the gullies. The depth of gully at Hesalong is controlled by the bed rocks.

On 25.7.78 we (the author and Shri A.K. Sarkar, Assistant soil Scientist, soil Convservation Dept., D.V.C. Hazaribag) observed gullies at Hesalong and Boothfirm. At Boothfirm about 1 Km. south of 'Jagriti Bihar' (a tectnical School under construction) the depth of the gully at head was 2.25 metres. The width was 3 metres. The gully was proceeding towards east.

At Hesalong the depth of the gully was 1.5 metres at head. The slopes at sides and head was vertical. "This striking vertically of the gully walls is due to cohesive nature of soil which does not crumble into gentle slope " (Ahmad, 1969 :VI).

It is a common feature of the gullies in the Basin that "vertical rocks develop on the uplands 2' - 3' (.6m. to 1 m.) away from the gully head parallel to the perimeter, the soil becomes loose by the rains and the whole column of it overfalls in the depression (Roy and Bhattacherya, 1967 : 5).

Expansion of Gullies :

The gullies are actively engaged in headward and lateral erosion. The following data show the headward expansion of the gullies at a terrific rate in different sub - catchments of the Damodar Basin :-

TABLE : 8 - I

Headward Expansion of Gullies

Location	Subcatchment.	Distance of gully head from ridge in 1950		Distance of gully head from ridge in 1965		Rate of increase per year		Slopes in %	Land use
		Feet	Metres	Feet	Metres	Feet	Metres		
Khawa	Usari	222	67.71	190	57.95	2.1	0.64	2	Fallow
Nimadih	Tilaiya -dam left	223	68.01	150	45.75	4.8	1.46	1	Fallow
Alardih	Bhera	515	157.07	450	137.25	4.3	1.31	2	Fallow
Dahu	Ranikhet	450	137.25	480	146.40	4.0	1.22	3	Occasionally Cultivated.
Sendur	Siwani	600	183.00	516	157.38	5.6	1.70	2	Fallow

Source : Roy & Bhattacherya, 1967 :5.

It is clear from the table : 8 - I that the headward expansion of gullies in the different sub - catchments of the Damodar Basin in Jharkhand varies from 2.1 ft/yr. (0.64 m/yr.) to 5.6 ft/yr. (1.7m./yr.). It is minimum in the Usari sub - catchment and the maximum in the Siwane Sub - catchment. This difference is due to local factors, e.g.

the nature of slope, the nature of waterflow, etc. Gorrie has noted that " a gain of 200' (61 metres) in new length for a single gully heads as between the survey record of 1929 and the air sheets of 1951 is not uncommon, but the really dangerous feature is the increase of the over - all perimeter of rotting bank, caused by the rapid widening of the trough along each recently cut gully (Gorrie, 1954). This is particularly observed in the area where a calcarious sub - soil is decaying faster than the top crust and forms potholes many metres back from the head of the active gully. Murthy (1970), from recent aerial survey maps of the Damodar Basin, has also observed recent gullies of 200' (61 metres) length which are totally absent on the old maps.

Distribution of Gullies :

It is obvious that gullies are more concentrated in the northen, central and westren parts of the Basin. The source regions of the Barakar, the Konar, the Siwani, the Haharo, the Ranikhet and the Deonad sub - catchments are severely gullied. It is minimum in Dhanbad area. Steep scarp zones are generally free from gullies due perhaps to thin layer of soil and exposed bed rocks. The western part of the Basin has been severely gullied. Here scarps are less steep.

Areal Extent :

Table : 8 - II shows that about 8.00 % of the total area of the Basin is under gully erosion. The Tilaiya Dam Catchment has the highest percentage (17.6%) of its total area under gully erosion. The Khudia sub - catchment has the lowest percentage (1.1%) of its total area under gully erosion. Maithon (10.4%). Tenughat (8.9 %) and Panchet (4.4%) have intermediate positions.

In the Konar catchment the percentage of its total area under gully erosion in the forest is the highest (5.9%) followed by Tilaiya (4.2%), Panchet (2.0%), maithon (1.8%) and Khudia (0.1%).

Figures in parent thesis indicate percentage over total land area.

Bihar Forest Department (Sen and others, 1967) conducted a sample survey in the dense forest of six sub - catchments. It has been estimated that an extent of about 20 % of the area, under forest in each sub - catchment, is under gully erosion. In the upper part of the

TABLE* : 8 - II

Statement of Erosion

Dam	Catchment in acres	Land area Land (in acres) eroded, eroding or subject to erosion						
		Sheet Erosion in Farm land	Sheet Erosion in Forest	Total Sheet Erosion land	Gully Erosion outside land	Gully Erosion Within	Total Gully Erosion Forest	Total Erosion thin forest
Tenu-ghat	1113750	173600 (15.6)	168100 (15.1)	341700 (30.7)	40150 (3.6)	58450 (5.3)	98650 (8.9)	440350 (39.6)
Panchet	1322850	237450 (18.0)	93350 (7.0)	330800 (25.0)	31550 (2.4)	26700 (2.0)	58250 (4.4)	389050 (29.4)
Konar	245000	42250 (17.2)	27950 (11.4)	70200 (28.6)	5400 (2.2)	14500 (5.1)	19850 (8.1)	90050 (36.7)
Tilaiya	194400	20500 (10.5)	44100 (22.7)	64600 (33.2)	26140 (13.4)	8200 (4.2)	34350 (17.6)	98950 (50.8)

Maithon	1312550	203050	164450	367500	112900	23800	136700	504200
		(15.4)	(12.5)	(27.9)	(8.6)	(1.8)	(10.4)	(38.3)
Khudia	145550	12250	4300	16550	1400	200	1600	18150
		(8.4)	(2.9)	(11.3)	(1.0)	(0.1)	(1.1)	(12.4)
GRAND	4334100	689100	502250	1191350	217550	131850	349400	1540750
TOTAL		(15.9)	(11.6)	(27.5)	(5.0)	(3.0)	(8.0)	(35.5)

* Source : Roy and Bhattacherya, 1967 :5.

Konar sub - catchment, spectacular gully erosion in the dense forest has been observed. Some of these gullies have reached their local divide, while others are actively engaged in headward erosion.

Genesis :

It is a complicated geomorphic problem to explain the genesis and distribution of gullies. As already noted, we do not see regular distribution of gullies in the area. In most cases gullies have developed along the finger tip stream but generally not along the whole course, in some other cases they are present along the middle course of the rivers. Some where one bank of the river is severely gullied while the other (opposite) bank is free from them.

In the light of the above - noted characteristics and distribution patterns we will examine the view related to its genesis :

Bennet (1955) and Brice (1966) are of the opinion that gullies are the result of unwise use of land. Bryan (1941) and Antevs (1962) think that they are due to climatic change. Emmett (1968) is also of the similar view when he says that "climatic change was responsible for the deterioration of the plant cover and initiation of gully erosion".

The opinions of these authors appear to be inadequate to explain the genesis of gulllies. It is true that these factors have important roles in the development of gullies. But the magnitude of gullies in the area cannot be explained either by unwise use of land or cyclic change of climate.

The genesis of gullies can be explained by the uplift of the area. The Damodar and its tributaries are rejuvenated streams. The Basin along with Chotanagpur persists to be rising (Dunn, 1942). The uplift accelerates the eroding capacity of channels. The high gradient causes rapid flow of water and gullies are formed due to channelling of water during rapid runoff.

Professor Ahmad, (1967, 1970) has presented a detailed study of gully erosion in India. He has indicated that these are limited only in four pockets, viz. the sub - Himalayan punjab, the Jamuna - Chambal region, Gujrat and Chotanagpur plateau region.

On the basis of his detailed study he has also concluded that " the continued uplift apppears to be the only probable explanation for the gullies". (Ahmad, 1969 : VI).

The regional uplift makes local conditions (e.g. the nature of slope, local relief or the height above the river or channel bed and runoff) favourable for the develoment of gullies.

8.7. Higher Rim :

It has been noted earlier that different planation surfaces of the Damodar Basin in Jharkhand descend to their immediate lower surfaces with steep scarps. But as we approach the edge of the planation surface, we have to go up prior to descending down .

Ahmad, (1965 : 381 - 382) is of the opinion that "the higher level of the belt is related to the original structure". The scarp zone appears to have an ancient folded structure. Numerous cases of high dips in the gneiss are seen. Examples of overfolds, recumbent folds, and contortions on small scales are observed in the scarp zones. The Higher rims above the general level of the planation surfaces may, therefore, be interpreted " as marking the eminences (stem of older mountains) of older erosion surfaces out of which the present surfaces and scarps have been carved" (Ahmad, 1976 : 55). The rims are extensive long narrow belts in the Dalma Region; Kodarma edge; Chutupalu area, northern, southern and eastern edges of the Upper Hazaribag plateau.

8.8. River Capture :

River capture occurs when " an actively low level stream encroaches on the drainage of nearby stream flowing at a higher level and diverts part of the water of higher stream" (Lauder, 1968 : 1054). The most common cause of stream piracy is the ability of one stream to maintain and extend its valley at a lower level than that of adjacent stream. It may take place by `abstraction' (Gilbert, 1877), headward erosion, lateral planation and subterranean diversion (Thornbury, 1954) Numerous sites of river capture may be observed in the area under study.

It has been noted earlier that the Damodar Basin in Jharkhand is a rejuvenated and re-excavated feature. In the early stage of the cycle of erosion when insequent streams form a fine - textured than others, and the divides shift laterally to engulf adjacent streams. In this way master streams are developed. This process, as may be observed, is still in progress, in the belt of severe gully erosion.

An excellent example of the river capture may be observed in the north - western part of the Basin. The bend in the Barakar river near Barhi has the appearance of an elbow of capture. There is no visible physical obstruction down the bend which would have deflected the river course. It appears that the Barakar in course of its headward extension captured a north flowing river in the neighbourhood of Barhi. The beheaded river is perhaps the Dhadhar nala a tributary of the Ganga. A recent upwarp along the Kodarma - Chauparan line has helped the wind - gap site to be eroded fast (Dunn, 1942 : 8- 9 ; Prasad, 1979 and Sharan, 1958).

Another notable example of river capture may be observed in the Siwani sub - catchment. The general slope of the area is towards north. The westward extension of the Siwani is perhaps, due to its headward erosion. In the process, it has diverted or captured the water of several streams flowing towards north directed by the

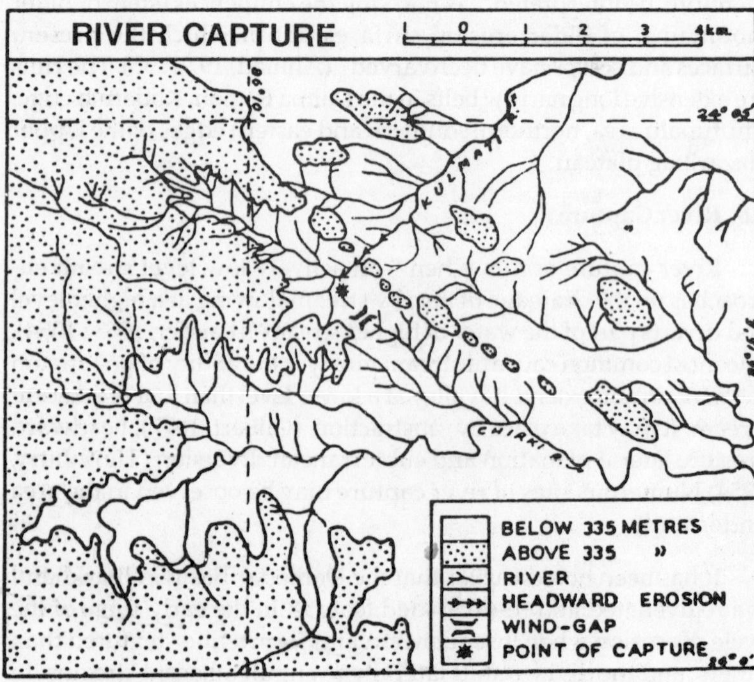

Fig. 21

general slope of the area. This has been evidenced that the beheaded streams are more or less on the same line.The divide between the tributaries of the Barakar and the Siwani is on the northerly sloping surface.

The tributaries of the Saphi and the Nalkari are constantly extending their valleys southwards by headward erosion and thereby pushing back the southern watershed far into the Ranchi planation surface and also show some evidences of stream piracy.

Fig. 21 shows a case of stream piracy in the Barakar Basin. The Kulmati river, a tributary of the Barakar, has diverted part of the water of the Ghatiari, a tributary of the Chirki river. The following facts support this contention :

(i) There is a ridge from Bishubera to gola dabar. The Ghatiari river originated from the southern slope of this ridge and flowed towards the east to become the tributary of the Chirki river.

(ii) The Kulmati river has extended its valley southwards across the ridge by headward ersosion and captured the water of the upper part of the Ghatiari river.

(iii) The law of identical tributary direction (Sharma, 1968 : 43) on both sides of channel reflects the occurrence of river piracy.

(iv) The presence of an 'elbow of capture' and 'wind gap' near Ghatiari village are suggestive of river piracy.

(v) There is no physical obstruction to divert the water of the upper part of the Ghatiari river towards the south.

9

Surface History and Evolution of the Drainage System

9.1. Surface History :

The Damodar Basin in Jharkhand is an old oplycyclic region. The surface was levelled and relevelled since it came into existence. The whole surface including Chotanagpur Plateau was reduced more or less to the base level before the Himalayan earth movements took place during the Tertiary and Post-Tertiary times. It was known as the " Pre-Tertiary peneplain" (Dunn, 1942: 12) or" early Cainozoice planation surface" (King, 1962 : 326). At present we may observe a series of planation surfaces at different- elevations, bounded by steep scarps on which streams flow through gorges, rapids and waterfalls, etc.

Thus, the present erosion surfaces of the Basin are a clue to the past. The Basin alongwith Chotanagpur Plateau was uplifted successively and intermittently due to the side effects of the Himalayan orogency. Hence, they are of Tertiary age and once formed the margin of the great sedimentary basin of Bengal, Assam and Arakan Coast, (Evans, 1971 387).

Regarding the causes of this rejuvenation, Dunn, (1939: 142) has explained the mechanism of subsidence of the Indo-Gangetic through and rising of the Himalayas and the Deccan Foreland. Wadia, on the basis of the studies of Himalayas has, Postulated three intermittent uplifts of Himalayas. The uplift of the Himalayan chanin was initiated during the Cretaceous and continued up to the Pleistocene (Mithal, 1968 : 67). Dunn (op. cit.) is of the opinion that the intermittent uplift of the Himalayas has corresponding periods of rejuvenaion in the Chotanagpur, which may be correlated to the interruptions in cycle of erosion. E. Ahmed (1958 : 24) seems to agree with this view when he writes, "as a side effect of the violent

Himalayan movements, parts of Peninsular upland in general and Chotanagpur peneplanin in particular began to be successively uplifted. "Hence, the present planation surffaces in the Basin are the parts of the same Pre-Tertiary Paneplain successively uplifted during the Himalayan orogeny.

There are four planation surfaces separated by steep scarps in the Basin :

Plantation Surface	Elevation	Age
1. Pat Region	1000m.	Eocene-Oligocene
2. The Ranchi and the upper Hazaribagh sufaces	600-700m.	Mid-Miocene
3. The Lower Hazaribagh or Kodarma-Giridih areas	100-305m.	Pleistocene
4. The broken Country of the Dhanbad area	100-305 m.	Pleistocene
5. The Damodar Trough	100-457 m.	Rejuvenated and re-excavated features.

9.1.1. Pat Region :

The Khamarpat (1068m.) area is the extension of the main Pat region. It is about 250 metres above the Ranchi Planation surface and more the 400 metres high from the Damodar trough. It is much dissected due to intensive fluvial erosion and mass wasting. Its flat-topped summit is accordant with the flat topped summit of the main Pat region. It is made up of outliers of Deccan Trap. The upper surface is convered with Laterite cap derived by the weathering of basalt.

9.1.2. Ranchi Planation surface and Hazaribag Surface :

(i) Avery small area of the Ranchi Planation surface is within the perimeter of the Basim. The surface is made up of Archaeans. It has an elevation of about 650 metres. The northern face of the plateau descends abruptly to the Damodar trough. The Plateau surface has low gradients. But in the scarp zone, high gradient (1in 4 or even more) has been observed especially in Chutupalu area. The streams are engaged in pushing the perimeter of the Basin towards south by headward erosion.

(ii) The Upper Hazaribag planation surface (610 m.) is seperated from the Ranchi planation surface (650 m.) by the Damodar trough (300-450 m.) The Chandwa- Tori- Balumath ridge (600 m.) of Archaeans connects these two surfaces. The area is composed of Archaeans. It steps down to the Lower Hazaribag Plateau with steep scarp. On the south the surface decends abruptly to the Damodar trough. But, the southern scarp of the Hazribag plateau is not as steep as the northern scarp of the Ranchi Plateau.

9.1.3. Kodarma-Giridih Planation Surface :

The Kodarma-Giridih or the Lower Hazaribag planation surface (350 m.) is occupied by a great variety of rocks. Isolated hills are common. Some of these can be identified as relics of the higher plateaus, but others, such as Parasnath (1368m.), the highest hill in Jharkhand, are relics of some much older surface. The height of the hills decreases towards east. This indicates the earliest slopes of the area before rejuvenation took place. The hills are Mahuda hill - (735m.), Mahabar hill (687m.), Gordhar (667m.) and Arwara (621m.), hills etc.

The planation surface (the Ranchi, the Upper Hazaribag and the Kodarma - Giridih) "resemble each other in many respects, that is, they appear to be flat, extremely gullied, studded with low rounded hills and bordered by steep escarpments" (Chatterji, 1969: 7).

9.1.4. Degradational Upland or Broken Country of Dhanbad Area:

In the eastern portion of the Basin the erosion surface is ununiform, formed by the river valleys, plains and isolated small hills locally known as tongaris. The elevation of the surface ranges from 100 m. to 305 m. The area is composed of a variety of rocks, e.g. gneisses and granites, schists, sandstones and shales, etc.

This surface is the product of the final uplift of the area, which was of the nature of a gentletilt in the east. It is because of the fact that there are no steep scarps on the eastern edge of the degradational surface of the Basin towards Dhanbad area.

9.1.5. The Damodar Trough :

The Gondwana trough, through which the Damodar flows, is

a rejuvenated and re-excavated feature. The presence of incised meanders, steep scarps, waterfalls and breaks in the long profiles of the rivers justify this contention. It is important to note that the trough area is more hilly than the adjacent plateau surfaces as a result of differential erosion of Gondwana rocks of varying hardness.

The northern and the southern boundaries of the trough are roughly determined by fault-line scarps. Post-Gondwana earth movements are responsible for the arching of Gondwana sedi-ments, thereby, accentuating the hilly character of the trough.

The width of the trough is increasing due to the recession of the scarps on either side of the trough as fluvial processes are active. This may be observed conspicuously in the Basin.

9.2. Drainage Evolution :

It is a complicated geomorphic problem to trace the history and evolution of the drainage system of the Damodar as the Basin in Jharkhand is "an old polycyclic area" (Ahmad, 1965 : 103). Several tectonic movements have complicated the matter. What we know about the evolution of the Damodar drainage system is on the basis of the geology and tectonic history of the region.

Geologically, the area came into existence in the Precambrian or Archaean era. So, at that time there might have developed drainage in sympathy with the folds (Singh, 1958 : 26)

The tectonic framework of the "peninsular craton was largely stablized by the end of the Precambrian tectonic cycles. Several well-marked structurally weak zones were, however, present within the craton" (Chatterji, et. al. 1970 : 10). The Damodar area is one of these weak zones. The development of the rivers might have been controlled by these weak zones.

From Cambrian to Permocarboniferous the area did not experience any earth movements of prominence. The denudational work might have reduced the area to a plane surface and rivers might have been flowing in senile meandering courses with graded profiles.

The Permocarboniferous rift valley played a vital role in localizing the regional dirainage into Gondwana trough. The rivers might have adjusted completely with the faulted basins and fracture zones, which " radiated to the west, north - west and north form the

Raniganj area" (Krishnan, 1953 : 22). The Damodar itself is probably a fault guided stream.

The glacial period of early Gondwana had also played a major part in the Basin. The ice appears to have covered the surface of the area and drifted towards the east. The movement of ice planed off the hills and filled the valleys with debris, thereby obliterating the former lines of drainage (Sharan, 1958 : 131).

The whole area might have been converted into lakes and swamps of freshwater due to melting of the ice, through which "the parent Damodar" (S.C. Chatterji, 1946) sometimes flowing with" great loads and at other times almost lost in stagnant water wended their way eastwards to sea" (Fox 1930: 20 - 30).

Gondwana strata are in fact, all of freshwater origin, laid down in large open flood plains, inlald lakes or swamps. Gee (1932) is of the opinion that" at least during the early part of the Damuda period. the general direction of the drainage of these low lying swamps and lacustrine tracts was from east to west". He had observed heavier inorganic rock materials deposited in the eastern portion of the fiels and the lighter substance of the coal carried futher west into the middle and western Raniganj and the more distant Jharia area. Probably, Gee might have been thinking about the tributary of the main stream. as he could not mention the presence of finer materials further west of Jharia coal field.

In the Talchir boulder beds of the Jharia Coal field, quartzite pebbles are found. These quartzite pebbles resemble the kaimpur quartzites. Fox deduced that Kaimur quartzites may be the source of these pebbles. If his view is taken to be correct, in that case there might be a line of drainage from Kaimur to the eastern part of the Damodar Basin.

The Gonwana trough developed in a line from Hutar to the Raniganj Coal fields. It is probable that this trough localized the Auranga - Damodar river, whose source may be supposed west of Hutar Coal field. Sharan (1958 : 134) opines that there was a tributary of this great system, from Daltonganj area.

M.S. Krishnan (1949 : 433) is, perhaps, also of the similar view when he says that from the general distribution of the Gondwana rocks it is now thought that there was one line of drainage to the north - east and east in the eastern part of the Damodar valley area.

This river probably crossed the present western plateau through Tori gap which is only 1600' (480 meters) in elevation (Sharan, 1958 : 134). The General elevation on either side of the gap is more than 900 meters.

The presence of Gondwana outliers, on the higher surface perhaps indicate the palaeoposition of rivers.

The rivers were engaged in severe erosion. The advent of monsoon climate during the Cretaceous time gave further impetus to their erosion action (Sharan, 1958 : 138). In course of time the Gondwana cover was removed away from all over the Basin expept in the faulted troughs and the drainage lines were superimposed on the crystalline basement. Deplanation progressed unawaited and by the end of Cretaceous, a widespread plane surface was produced. This plane was called 'Pre-Tertiary peneplain' (Dunn, 1942 : 12). This may also be called as the Cretaceous planation surface on which there could be "no scarps or waterfalls," (Ahmad, 1965:24)

One of the major events was " the vast lava outpour about early Eocene which no doubt burried the pre- exiting drainage lines in the area of lava outpour" (Ahmad, 1965 : 10). This erruption actually played a very significant part as the central portion of the Auranga - Damodar river was buried. Numerous consequent Streams originated from the newly formed surface. The source region of the present Damodar is the Khampur Pat (1068 m.). The Tertiary orogeny did not alter the major trends of the streams of the Basin. But we may say that it has given a new life to the rivers. The Damodar in Jharkhand is a rejuvenated and re-excavated feature.

Thus, we may conclude that the present drainage system of the Damodar is the result of the rejuvenation of the area during the Himalayan storm. The presence of the scarps, the incised meanders, the waterfalls, rapids and the terraces probaly support this concept.

It has been proposed by D.V.C. (Indian Nation, Patna, 17.7.1978) to change the course of the Damodar in the Bokaro Coal field area. If it is done, it will be a manmade change in the course of the Damodar river.

10

Models of Landscape Evolution

The morphological variations of the landscape are not accidental or haphazard. Their patterns and processes can be explained to a great extent with scientific precision. Whether various landforms in the area have evolve in 'evolutionary sequence' or they are 'non - cyclic' in origin, are the complex basic geomorphological problems. Many geomorphologists have contributed to the problem. Hence, a brief review of the views and works of different authorities will go a long way in solving and giving proper explanations of the methods. models and processes involved in the evolution of the landforms of the Damodar Basin in Jharkhand.

10.1. Davis Model :

Before Davis, Powell (1875), Gilbert (1877), Dutton (1882), and several others contributed to the field of geomorphology. Powell introduced the concept of 'base - level', the level below which the land cannot be eroded. Dutton (1882) gave greater precision on the base level idea when he wrote " all regions are tending to base levels of erosion and if time be allowed enough, each region will, in its turn approach nearer and nearer and at last sensibly reach it.".

Davis, W.M. (1899, 1909) synthesized the concepts of earlier workers and moulded them into a unified system for the geomorphic interpretation of landforms. The entire system of the landscape evolution was called as the 'geographical cycle', the end product of which was supposed to be a surface of low relief or 'peneplain'. The Davisian peneplain is due to lowering of divides over the periods of prolonged tectonic rest. But lowering of divides may also occur where rate of uplift is less than the rate of erosion (Pesci, 1970 : 18).

Davis states that all the varied forms of the lands are controlled by geologic structure, geomorphic processes and stages of development. The concept of geographical cycle of Davis (according to system theory) is recognised as an example of closed system (Harris & Twidale, 1968 : 237). His model suggests a sequential development of landscape through stages of youth, maturity and old age, with general slope decline ultimately producing a peneplain. Admittedly the whole cycle includes everything from plain to plateau or mountainous upland and back again from upland to plain (A.Holmes, 1975; Chorley, et al; 1964, 1973; Small, 1972; Von Engeln, 1960; Thornbury, 1954; Twidale, 1976; Rice, 1977; Wooldridge and Morgan, 1959 ; Sparks, 1972 and Leopold, et al. 1964).

10.2. Penck -King Model :

Penck (1924) model provides a genetic study of landforms, the geometry of slopes and their relation with structural movements. He regarded surface morphology as the result of 'competition between crustal movement and denudational processes' (Rice, 1977 : 216). He argued that the primary control of landforms evolution is the relationship between the rate of river downcutting and the rate of tectonic uplift.

Penck's valuable contribution lies in the sphere of the development of slopes. The idea of parallel retreat of slope was conceived by Fisher and Penck. Bryan (1940) also contributed to this problem. This concept has been developed into a thoroughgoing hypothesis of the landscape development by King (1953, 1962). He has argued that the world's plain lands are not peneplains but pediplains formed by stream incision, scarpretreat and pedimentation and that they consist of coalesced pediments.

10.3. Hack Model :

Hack was a strong exponent of 'noncyclic' approach to landform studies. His model suggests that landforms are the product of competition between the resistance of crustal materials and the forces of denudation (Rice, 1977 : 216). His theory of 'dynamic equilibrum' is not new. As early as 1877, Gilbert applied it in the Henry mountain. Nikiferrof (1942) made similar proposals with respect to pedology. Davis (1922) was well aware of dynamic interpretation between active crust and external agencies on the one hand, and likelihood of self - regulatory processes on the other. Prior to Hack, Strahler (1950, 1952) also applied this principle in geomorphic interpretation of slopes, etc.

The concept of dynamic equilibrium "requires a state of balance between opposing forces such that they operate equal rates and their effects cancel each other to produce a steady state, in which energy is continually entering and leaving the system" (Hack, 1960 : 82). Hence, (according to system theory) it is the example of `an open system' (Chorley, 1962).

10.4. Discussion And Conclusion :

In the light of the above models we have to discuss geomorphic features with a view to see as to which of the above models is applicable in the geomorphic interpretation of the landforms of the Damodar Basin in Jharkhand.

(i) The landscape of the Damodar Basin in Jharkhand bears the imprint of intermittent and successive uplift. The area has been uplifted epeirogenetically during the Himalayan orogeny. This may be evidenced from the presence of four planation surfaces e.g. the Khamar Pat 1000 m., the Ranchi and the Upper Hazaribag planation surfaces 600 m., the lower Hazaribag planation surface 300 m. These planation surfaces descend to their immediate lower surface with steep scarps.

If we mould the evolution of landscape into Davisian model, it needs a prolonged period of stability (tectonic rest) during which the landscape evolves towards a surface of low relief related to the contemporaneous base level. It means every uplift was followed by progressive denudation of land surface eventually producing peneplain. It may also be possible that uplift might have taken place before the completion of the cycle of erosion. The last upheaval took place during the middle of pleistocene period. The pleistocene period dates back 2-3 million years.

The average height of the Damodar Basin in Jharkhand is appproximately about 733 meters (taking the averages of the sum of the highest point e.g. the Parasnath Hill 1366 m. and the lowest point e.g. at the confluence of the Barakar with the Damodar river 100 m.). A loss of 7148 cu.m./Km.2/year has been reported by Murphy (1970). It means the rate of denudation in the Damodar catchment is a little over 7 mm./ year, ignoring the fact that it will vary from one part of the Basin to another, due to several physical factors. Thus, a little over 100000 years are, probably, sufficient to reduce the landscape near the base level or the 'peneplain' of Davis.

If it is taken to be correct the whole Basin should have reached the base level long before. The valid explanation of Davis model may be that the area might have attained many more times higher surface than the present level. This goes against the concepts of Dunn (1942), Ahmad (1958) and Singh (1969), who have worked in the area. It is true that the area must have suffered deplanation or degradation. But the rate of past degradation is not known. It is difficult to trace the actual elevation of the surface during its uplift. Dunn assumed successive uplifts of the different segments of the Pre-Tertiary Peneplain by faulting or warping along or near the present lines of scarps. But this concepts faces the difficulty in visualising the faults along the scarps.

Ahmad, (1958, 1969, 1976) has advanced the idea of the regional uplift and parallel retreat of scarps in geomorphic interpretation of different planation surfaces of Chotanagpur plateau. He assumed the successive regional uplift of Chotanagpur (the Damodar Basin in Jharkhand is a part thereof) with reference to the surrounding base level (namely the Ganga Valley). Scarps, which bound the planation surfaces, aquired their present position merely because of recession.

The study of stream-line surfaces (Chapter :6.6) of the area strongly supports the view of Ahmad. The study reveals that the lower planation surface is expanding its area, while the higher planation surface is decreasing in its areal extent by recession. In the process of slope retreat several hills (e.g. Rochap and Mahabar, etc.) have been left isolated. These hills, more or less, match in the elevation with the scarps. Field examples of the retreat of the valley-side slopes may be observed at Chandro in the konar valley. Regional uplift of Chotanagpur also explains the question of isostasy. This concept of Ahmad suppports Penck - King model of the landscape evolution, which relates the forms of the landscape to changes in the rate of uplift relative to the rate of erosion. Parallel retreat of slopes results in pediments and they coalesce into pediplains. Whether the landscape of the Damodar Basin is a 'peneplain' or pediplain' is an open question ? It requires an intensive field work.

One remarkable point to be noted is that on the planation surfaces of the Basin e.g. from Chandrapura to Jarangdih and also in other parts of the area, the undulations are more marked and the slope of the planation ssurface between adjacent streams increase.

This probably does not fit in a scarp -retreat -theory of frontal pedimentation where the eminences should be trimmed off. It also, probably, is not in line with degradation and downwasting of divides implicit in Davisian concept of peneplanation. Such a downwasting be universal unless there is regional or local upwarping or uplift leading to comparative unevenness visible in the belt under reference (Ahmad, 1970: 10).

'Dynamic equilibrium' of Hack may explain the landscape. It suggests that "every geometric elements is the product of adjustment between materials and process" (Tanner, 1968:316). Differences in geometry are to be explained in terms of local variation of materials and processes etc.

(ii) "Fluvially-eroded terrains are, in essense assemblages of hillslopes. An appreciation of the mechanisms responsible for shaping the slopes is, therefore, critical to an understanding of the landscape as a whole" (Rice, 1977 : 139). The sample study of the hills reveals that they are asymmetric in slopes i.e., the slope of one side of the hill differs from another . Again measured profiles of the hillside slopes are not uniform. A casual look at the slopes shows their forms as convex crest, rectilinear, debris slope and concave bottom slope. These are the essential elements of the hill slopes of Wood (1942) and King (1953). But measured profiles are characterised by convexity, concavity and rectilinear units. Thus they are of composite nature (Prasad, et al. 1980). They may be called composite slopes with a convex crest, a straight line middle slope, and a concave slopes at base in general. But a close observation of profiles reveals that these units do not seem to follow this general sequence, rather they have been developed `non - sequential: These observations create a problem for the parallel retreat of hillslopes of Fisher, Penck, King and Brayan, and slope reduction of Davis. Hack model of the landscape development may explain these characteristics of hill slopes, because energy applied to the landscape evolution may vary from one side of hill to another depending on the volume and head of water.

Again it is easy to mould 'composite' nature of slope into the Hack model of the landscape evolution. According to Strahler's' euqilibrium - slope hypothesis' (which is more or less similar to slope system of dynamic equilibrium of Hack), the slope angle represents a condition of balance involving not only rock type and structure, but also climate, soil, vegetation and other circumctances. If one of these conditions was to change, then we could expect the angle of equilibrium to change as well.

(iii) The study of the drainage divides (Chapter :6.8) reveals that they have been migrating towards the gentler slopes. Following the 'law of unequal slopes' the streams (e.g. the Saphi and the Nalkari, etc.) on the steeper side are engaged in pushing the divide back in to the gentler slopes.

The migration of divide may be moulded into the Penck- King model of the landscape evolution i.e. the retreat of the slopes. *Fig.* 25 shows that they are not uniformly spaced. Asymmetric nature of the slopes on the two sides of the divide is perhaps respondible for the differential erosion and migration of the crestline. The stage of development on both the sides is, probably, the same. Thus, it may be explained as due to variation in energy on the two sides of the divide, which is the essential principle of the Hack model of the landscape evolution.

(iv) The existence of 'Higher Rim' (discussed earlier) on the scarp zone does not seem to support Davisian concept of the landscape evolution. Because according to his principle of slope reduction it should have been denuded and removed earlier. This may be interpreted by Penck - King model of the landscape evolution i.e. the slope retreat and the lengthy survival of old surface.

"For the existance of this feature there may be two alternatives at least. Firstly, the force of reduction or retreat did not act upon, and secondly they have acted upon the surface, but the force of dynamic equilibrium has given the maintenance of this feature; although there may be a case that the older stems of mountains do not exist in the original form, rather represent the forms under the new `environmental conditions'. The second alternative is more logical than the former, and can be more plausibly explained by a state of dynamic equilibrium which changes with the change in environment" (Prasad, 1980 : 94).

(v) The pattern of the drainage has been greatly influenced by the geological structure and tectonic evolution of the area. The river systems show high degree of adjustment to the structure. It means that they flow and have excavated valleys along lines of weakness in the underlying bed rocks. They are also closely adjusted to faults and fractures, joints, cleavage and bedding planes. This may, probably, indicate that the region has passed through more than one cycle. These strongly support the view that the Basin is 'polycyclic' region. The 'valley - in - valley' at Rajrappa also supports this contention.

(vi) The study reveals that drainage texture is indifferent to the stages of the landscape evolution. The areas of high relief, e.g. hills, ridges and scarps have generally coarse to medium texture. The fine and ultrafine textures are seen in a few limited pockets on the erosion surfaces of relatively softer rocks and also to some extent along the foot of the scarps, hills and ridges. The texture of drainage net is obviously related inter alia to the geological structure of the area.

Davis model does not seem to explain this observation because fine and ultrafine texture of drainage should be associated with the earlier stages of cycle of erosion. In the later stage integration of streams or absorption of minor streams by major streams may result in coarse texture. But here we see that the matter has been complicated (mainly tectonic uplifts). This appears to be explained by 'dynamic equilibrium' hypothesis of Hack. According to Hack landforms are the result of the "competition between the structureal features and environmental conditions of area" (Rice, 1977 : 216).

Similar is the case with the distribution and density of the hydrographic networks i.e. the source heads and the confluence points and explanations are more or less the same as noted earlier.

The above discussions conclude that the morphological variations of the Basin are neither related to slope decline of Davis nor slope retreat of Penck - King in all aspects. Both these phenomena are partially observed. They are most probably related to 'dynamic equilibrium'. Examples of dynamic equilibrium include asymmetric valley - side and hill - side slopes, composite nature of hill slopes, incised meanders, migration of divides towards the gentler slopes, high density of hydrographic networks (source heads, and confluence points) and fine texture of drainage relatively on low or moderate slopes, etc. The major changes in the evolution of landscape took place during the successive Tertiary uplift and will probably, take place after an initiation of a new set of environmental conditions (tectonic, climatic or volcanic, etc.).

Now the question arises as to how it is 'noncyclic'. Hack claims that landforms adapt rapidly to changing environmental conditions. This change in environmental conditions requires time which must have development phases. It carries the implication of sequential development phases. It carries the implication of sequential development though it may differ from Davis (Youth, maturity and old age) or Penck (aufsteigende, gleichformige and absteigende). The term 'new environmental condition' of Hack is suggestive of an 'initial surface' (Davis) or 'primarrumpf' (penck). Similarly, before the commencement of a new environmental condition in the Hack model, the stage is 'endrumpf' (Penck) or the 'surface of low relief' at the end of the Davisian cycle of erosion. It may be like peneplain of Davis or 'fastebene' of Penck. Hack did not relate peneplains to the erosional cycle, but he asserts that they are representatives of dunuded surfaces formed in a certain environmental condition, i.e. the surface of homogenous structure, drainage and slope. So it is a difference only in terminology rather than in concept (Prasad, et. al. 1980 : 98). Thus, inspite of the objections against the sequential development it is probably difficult to give up the cyclic concept.

From the above analysis" it seems that it may be necessary to seek some compromise between the over - simple cycle concept as outlined by Davis and the agnostic attitude adopted by Hack" (Rice, 1977 : 223).

Bretz (1962) and Holmes (1964) have strongly contested the view of Hack. But the morphological variations of the Basin may best be explained by the 'dynamic equilibrium' with the stages of cyclic order.

11
Climatic Geomorphology

11.1. Climatic Geomorphology :

• Climatic Geomorphology studies the evolution and development of landforms in any part of the earth in terms of climate. Accoding to this school of thought climatic parameters directly or indirectly control the genesis of the landscape. The works of Davis on normal cycle, arid cycle and glacial cycle give the idea of the climatic geomorphology. He, not only, recognised that distinctive assemblages of geomorphic processes and landforims occur in areas with specific climates in-different part of the world but also (1904), that climatic accidents could occur in any given region, resulting in the formation of the polygenetic landforms (Stoddart, 1969 : 161).

11.2. Climatogenetic Geomorphology :

Climatogenetic geomorphology is the historical analysis of climatic geomorphology. Time is a factor, which must be taken into account because "existing relief is not everywhere merely a synthesis of present structural features, but contains also older relief generations originating in earlier period of climatic and tectonic history. The investigation of these varied generations in the evolution of the relief is the task of climatogenetic geomorphology" (Budel, 1968 : 417).

Budel (1970) is perhaps, the strongest exponent of climatic and climatogenetic geomorphology. He distinguished these two concepts as "climatic geomorphology sensu - stricto is largely systematic study, and climatogentic geomorphology becomes the his torical analysis of particular area" (Stoddart, 1969 : 163).

Numerous scholars e.g. Budel, 1959, 1968, 1973; Bryan, et al. 1943; Carlston, 1966; Chorley, 1957; Cotton, 1969, 1973; Martonee 1973; Penck, 1965; Pesci, 1968; Pesci and Szilard, 1968; Stoddart,

1969; Tricart, 1972, 1973 and several others have emphasized the influence of climate on morphogenetic processes and associated landforms assemblages.

They put forward the following facts :

(i) It is assumed that in certain cases climate sets in motion isostatic reaction. For this glaciation and deglaciation are to be noted. Deglaciation provides the isostatic imbalance and hence the uplift of the area. Similarly, glaciation may be responsible for the subsidence of the area, "climate in this case has tectodynamic repercussions, particularly on the base level and general geomorphological evolution" (Tricart, 1974 : 14)

(ii) "Since climate has demonstrably changed during the Tertiary and Pleistocene, and continues to change, sets of climati- cally - controlled landform features are being continuously su-perimposed on each other, so that the landforms of any one area bear the imprint of climates no longer operative" (Stoddart, 1969: 163).

Thus "the present - day landforms are for the most part not the result of the present climate but the produce of Pleistocene processes. Consequently the present day active forces and their morphological expresseion depend on the climatic zones in which they occur" (Passarge, 1973 : 91)

(iii) It is claimed that the significance of pertrovariance has been overestimated by the structural geomorphologists. The ability of rocks to withstand erosion does not depend on their physical characteristics only as per Budel. He specifies the importance of climatic variation forgetting the fact that without tectonic activity no surface can be subjected to erosional activity. "Granite, for example, is one of the hardest rocks in the mid latitudes whereas in the moist tropics, where deep weathering prevails, it is soft and sand - like" (Budel, 1973 : 203).

(iv) They claim that epeirovariance is not, as a whole, a determining factor of the relief features. It governs the ultimate height and general steepness of slopes. Of course, "the forms of an uplited block or a step produced epeiro - genetically appears in its pure form only in the relief of the ocean bed on land, the effect of exogenic erosion processes begin at once, changing the raw - block into a greatly reduced, many - memberd relief of flat surfaces, ridges slopes and valleys which we see today" (Bide;, 1973 b : 204).

(v) King (1953) places process first in importance in the list of Davis's trilogy (structure, process and stage). From this it folllows that climate is one of the most important elements in determining the landscape type, as process is partially dependent on the character of the climate. Subsequently, however, he considered that the difference between the humid - temperate, semi - arid and arid landscapes are only a matter of degree (C.A.M. King, 1966 : 8).

Since climatic variance controls all dynamic processes, all over sturctural components, the climatic geomophology will become the decisive part of geomorphology of the future (Budel, 1970 : 30).

(vi) It is true that nearly all landscapes are polygenetic in nature at present time is so close in time to the colder periods of the Pleistocene. They show the influence of two or more climatic types; thus deserts show evidence of periglacial or glacial having been in operation. This complicates geomorpholgical interpretation because present forms are in many instances not adjusted to the process now acting on them. This gives weight to the views of the school of thought that emphases the importance of climatic elements in the landscape evolution (C.A.M. King, 1966 : 9).

11.3. Climatic Conditions and Paleoclimatic Characters of the Area:

The present climatic conditions of the Damodar Basin in Jharkhand have been note in the earlier chapter (Chapter : 6.3). Here it is essential to note that the climate of the area is subtropical mon-soon with alternate dry and wet seasons. The distribution of rainfall and temperature varies considerably in the area.

No direct records of past climate are possible. What we trace is based on the geological formations in the area which provide the indirect evidence for the climatic vicissitudes.

The planation surfaces of the Damodar Basin in Jharkhand have dearth of the records of past climatic conditions. But the formation of Dharwarian rocks (e.g. in Kodarma areas) suggest their subaerial sedimentation in primeval depressed zones, probably due to denudation of the higher surface "in dry, arid and semi-arid climatic conditions with phases of sedimentation" (Kumar, et al; 1980 : 353).

The Gondwana formation shows the evidence of several climatic changes the Carboniferous period to the Jurassic period. It began with cold climate about 275 million years ago. Examples may include the Talchir Boulder beds. Evidence shows that glacial climate was extended over the whole of the Peninsular India from Rajasthan to the Eastern Ghat and also over the Southern continents e.g. Australia and Africa. In the later part of the Talchir period or towards the close of the carboniferous period the glacial climate was changed to favour a vast amout of fluvio - glacial sediments and varved clays over boulders. "The prevailing greenish tints of the rocks indicate that the climate was still quite cold. Though there are plant impression in the beds immediately overlying the boulder beds, the flora was apparently still scanty during the Talchir period" (Krishnan, 1956 : 304).

Thick coal seams of the Damuda series probably suggest warmer and more humid climate during the permian period (About 215 million years ago) which permitted the growth of luxuriant vegetation. The strata are composed of sandstones, shales and coal seams in successtion. "The succession is repeated many times land during the whole of the Damuda Period there may have been as many as 50 or 60 cycles of sedimentaion" (Krishnan, 1956 : 305).

This was followed by another climatic change during Triassic Period (about 180 Million years ago). It is indicated by a slight unconformity between Raniganj (upper permian) and Panchet (Lower Triassic) beds. Absence of Carbonaceous matter in Panchet beds suggests that sediments were laid down in flood plains and shallow lakes." The presence of Labrynthodonts in them indicates a period of gradual drying up of the sedimentary basins" (Krishnan, 1956 : 306).

During the Middle and Upper Triassic Period the climatic conditions remained arid. Mahadeva series is composed of ferugenous sandstones often with thin layer of hematite and are entirely devoid of carbonaceous matter these seem to support the arid climatic condition. "The Glopssoteris flora parctically died out during the dry period of triassic" (Krishnan, 1956 : 306).

With the onset of Jurassic Period the climate was changed from arid to moist. Relatively humid climatic condition seems to have continued with perhaps minor variation during the rest of the Gowdwana era.

During the Tertiary and Post - Tertiary Periods the entire Chotanagpur Plateau (the Damodar Basin in Jharkhand is a part thereof) experienced epeirogenetic and successive uplifts. The climate changed form cold to warm and humid from Eocene to Oligocene and again to cold in Miocene - Period. No sediments of these pe-riods are seen in the area probably due to uplift and consequent rejuvenetion in degradational processes.

Laterite is "the result of the subaerial decomposition in situ of basalt and other aluminous rocks under a warm humid and monsoonic climate" (Wadia,- 1975 : 378). It is obvious that with the advent of the Monsoon climate the processes of laterization began and is active even at the present day.

It is important to note that the climatic fluctuation during Pleistocene was related to glacial advances in the Himalayan region. No trace of Pleistocene glacialtion is available in the area. But it is almost certain that each glacial maxima corresponds with a period of increased seasonal humidity (fluvials) in the Peninsular part (Kumar, et al. 1980).

The presence of some Himalayan plants particularly rhododendon in the high hills of peninsular India and in Parasnath hills (1366 m.), points out the possibility of Himalayan flora migration during glacial ages. It is almost certain that during the Ice Ages, the Peninsular Parts experienced a mild temperate climate (Kumar, et al. 1980).

The above discussions show that in the past climate changed consideraby. The present landscape of Chotanagpur is the out come of the last phase of the tectonic disturbance and subsequent denudational processes. "Although climatic conditions have varied in terms of distribution, the basic climatic types have not altered drastically in the last few million years "(Kumar, et al. 1980 : 358). Hence, the influence of past climate on the evolution of the present landscape cannot be the prime factor. It has a minor and secondary role to play. The primary part in the landscape evolution is that of tectonic movements and structure.

11.4. 'Fossil' Erosion Surface :

Most of the bed - rock surface of the area is covered with thick or thin layer of soils and sediments. The depth of sediments varies from less than a metre to more than seven metres on the planation surface. But the hills, ridges and scarps are generally rocky. It is the general feature of the whole of Chotanagpur Plateau where the bed - rocks are concealed under the mantle of waste and debris.

Eminent geomorphologists e.g. M. Pecsi, P. Macar C. Embleton and A. J. Wiggers during their field trip to Netarhat in connection with the symposium on 'Erosion Surface' in the dept. of Geography, Ranchi University, Ranchi, indicated that Chotanagpur Plateau is not an active erosion surface because sub -aerial processes are not active on bed - rocks . It is a 'fossil ero-sion surface' or 'dead erosion surface' which represents an earlier arid climate.

They made the following observations:

(i) On the planation surface of Chotanagpur, bedrocks are covered with the mantle of waste and debris. The upper surface of bed - rocks and the present ground are more or less parallel suggesting that the present ground surface has been inherited from the bed - rocks which was earlier at higher level than the present ground surface. Subsequent weathering and erosion of bed - rock have converted the upper layer into a mantle of waste and debris, (Prasad, 1979 : 322).

(ii) It is opined that the Chotanagpur Plateau was under arid climate in the geological past, which shaped the landscape. Now, the climate has changed into sub - tropical monsoon climate condition and causes an enormous production of waste and debris.

(iii) Pesci observed laterite, red clay and gravel beds on the side of a small river between Lohardaga and Ghaghra. The most important point was the occurrence of the gravel not merely as a features of the river bed but as a layer inserted in the vertical profile of the side of the stream and directly on bed - rock. Subsequently clay developed on the gravel bed. The gravel dates back to the earlier more arid times when Chotanagpur was and active erosion surface and the forces of weathering and erosion were directly active upon the bed - rock leading to the formation of the pediment floor and the accompanying gravels. Subseqently the climate became more humid resulting in a higher rate of weathering and waste production and the burial of the bed - rock by sediment. This phase is represented by the overlying bed of clay on which tropical conditions have produced laterite (Ahmad, 1970 : 12).

Prof. Pesci had observed "pediments and rock plains in a live erosion surface east of Mysore plateau, but on the Ranchi Plateau now, erosion has ceased to operate on bed - rock which lies bur-ied under a thick mantle of waste" (Ahmad, 1970 : 9).

The above observations of the foreign delegates conclude that the Chotanagpur Plateau is not an active erosion surface. It is a fossil erosion surface or dead erosion surface. They conclude that these features may be explained as due to change in climate from arid to humid (monsoon).

11.5. Tectonogenetic Geomorphology an Alternative View :

The claims of climatogenetic geomorphology has been noted earlier. But the significance of tectonogenetic geomorphology cannot be ignored in the geomorphological interpretation of the landscape evolution. It "refers to the tectonic activities of the past ages" (Ahmad, 1970 : 3). Landforms that result from crustal movements are described as tectonic. "The geosynclinal strata and any accompanying volcanic rocks were eventually folded and crumpled, faulted overthrust and more or less intensely metamorphosed i.e. subjected to a group of processes collectively known as tectogenesis" (Holmes, 1975 : 1109).

Ahamad (1958, 1969), Dunn (1942), Sing (1969), Prasad (1979, 80) and others have observed that the morphological evolution of Chotanagpur Plateau (the Damodar Basin in Jharkhand is a part thereof) has been controlled by the tectonic activities in the past. The present surfaces are the result of the latest phase of the lectonic activities. But the imprint of the Archaean orogeny may also be observed in the area. It includes the thrust folds, recumbent fold, eminences of old surface in the scarp zones' and banded geneiss, etc.

11.6. Significance of Tectonic Activities in the Area :

(i) Most of the area consists of granites and gneisses of Archaeans. Though the trough area does not consist of granites and gneisses on the surface, these rocks do occur concealed beneath the Gondwanas or superficial deposits. The area has remained as a land region from the dawn of the geological history. So, it must have undergone several planations and tectonisations.

Recumbent folds, overthrust, etc. clearly indicate that the area had been affected by the severe Dharwarian earth movements. The Vindhyan era saw the invasion of the sea on the north - western corner (outside the area under study) of Chotanagpur plateau and its subsequent uheaval.

The Permo - Carboniferous earth - movement played an important role in the physical history of th area. The region did not experience any compressional orogenic deformation. The area suffered from tension. Linear Cracks were developed. The Damodar trough came into exitence. Sediments were deposited in the trough area which led to the formation of thick deposits of several thousand metres. These sediments were subsequently in truded by dykes and sills.

Between the Permo - Carboniferous earth - movement and the Tertiary orogeny, there was tectonically silent period. The exogenetic forces during this long tectonically silent period was enough to reduce the area near to the base level on which there were no scarps, waterfalls, incised meanders and gullies, etc./

Tertiary and Post - Tertiary earthmovement has played a significant role in physical landscape of the area. The entire Chotanagpur plateau was uplifted successively and intermittently. The movements have given rise to different planation surfaces of area at different altitudes. The well - defined Damodar Valley is probably partly due to the uplift of the area and partly due to re-excavation of softer Gonwanas, on the same site as that of Permo-Carboniferous rift valley.

It is clear that the present landscape of the Basin is eminently and clearly the result of the epeirogenetic uplifts of the area during the Tertiary and Post - Tertiary tectonic movements.

(ii) During the seminar on Himalayan Geology held on Sept. 13 -17, 1976, many papers led to the line of thinking that "the Himalayan was built up out of the products of tectonisation of the frontal edges of the Indian Peninsula, sliced up as wedges, during the interaction of plates " (G.S.I. 1976 : 33). The theory of plate tectonics has been advanced to solve many of the geophysical, seismicity and structural features, etc." (Ahmad, 1978). So the Indian Peninsula is working as a plate which is moving northward. A recent exploration of Indian Ocean has revealed that the crest of the oceanic ridge is diverging outward due to tension at the rate of 5 to 8 Cm. per annum (Prasad, 1980). These phenomena apparently show the Peninsula is subject to tectonisation.

Thus, we see that tectonic activites played a primary-role in shaping the landscape of the Damodar Basin in Jharkhand. It is assumed that the Basin is still rising. We do not know the actual rate of uplift. But without some uplift or tectonic disturbance streams cannot indefinitely continue to erode the surface. The ungraded nature of streams, the headward erosion, and the magnitude of gullies may probaby be explained due to gradual or intermittent uplift of the area - keeping pace more or less with erosion.

11.7. Observation and Discussion :

It is obvious that the prominent supporting point in favour of the fossil erosion surface is the presence of the layer of gravels over the bed - rocks. This raises at least the following questions :- (i) is there any gravel bed overlying the bed - rock, and (ii) if the layer or gravel does exist, is it of drift or in situ in origin?

For this purpose several soil profiles in the different part of the Damodar Basin in Jharkhand have been examined. We do not find any gravel beds on the bed - rock. This fact has also been supported by the soil profiles prepared by the soil conservation Department of the D.V.C. (1972). A brief accout is given below:

(i) It is clear from the soil profiles that the layer of gravels does not occur on the bed - rock as suggested by Pesci, et al. Most of the soils or sediments are residual or in situ in origin. They are derived mainly from the bed - rock or parent material. Weath-ered or decomposed rocka may be observed at a depth of 7 metres or even more.

This idea that these soil or sediments in the area have been deposited from the surrounding higher surface does not seem to be acceptable, because we do not find the alternate layer of coarse and fine sediments in the soil profiles.

Soil Profile No. 3

Sub catchment	-	Usari.
Village	-	Maheshpur
Relief	-	Undulating upland
Slope	-	1-3 %
Drainage	-	Surface runoff rapid, internal drainage moderately slow
Vegetation	-	Occasionally cultivated for

millet crops such as Marua & Gondly, Maize is also grown.

Horizon	Depth in C.M.		Description
A_1	0-17	:	Very pale brown 10 YR, 7/3, (7.5 YR 5/4), Loamy sand, single grained, loose dry, friable moist, very rapidly permeable, plenty of roots, clear boundary, PH 5.4
A_2	17-30	:	Yellowish brown 10 YR 5/4 (7.5 YR 0.5), sandy loam, granular to very weak sub-angular blocky, slightly hard, friable moist, rapidly permeable, plenty of roots, diffused boundary, PH. 5.4
A_3	30-47	:	Strong brown 7.5 YR 5/6 (5.5), sandy clay loam, weak sub-angular blocky, slightly hard, sticky moist, moderately slow permeability, PH 5.3.
B_1	47-107.5	:	Yellowish red, 10 YR 7/6 (5 YR 6/6 BY), clay loam, sub - andgulay blocky, hard, sticky moist, slowly permeable few roots, PH 5.6.
C.	107	:	Weathered granite and gness.

Soil Profile No. 4

Sub - catchment-	-	Jamunia
Village	-	Dumari.
Relief	-	3-5%
Vegetation	-	Putush and Palash

Drainage		-	Moderately well drained, mod erately permeable

Horizon	Depth in C.M.		Description
A_1	0-23	:	Yellow 10 YR 7/6 (10 YR 5/6), loamy sand, single grained, loose dry, loose moist, rapidly permeable, plenty of roots, clear boundary, P.H. 5.0
A_2	23-37	:	Brownish yellow, 10 YR 6/8 (10 YR 5/6), Sandy loam, medium weak sub-angular blocky, slightly hard dry, very friable moist, plenty of roots, deffused boundary.
B_1	37-58	:	Yellowish Brown 10 YR 5/8 sandy clay loam, sub-angulay blocky, hard dry, firm moist, sticky wet, clear boundary, few roots PH. 7.0
C	58+	:	Weathered Quartz vein and Feldspar.

Soil Profile No. 5

Sub- catchment	-	Gowai.
Relief	-	Very gently sloping upland.
Parent material	-	Granit gneiss.
Vegetation	-	Grass, Pipal Ber, Palash.
Slope	-	1 - 3%
Erosion	-	Moderate sheets.
Location	-	Village : Talagaria, Dist. Dhanbad

Horizon	Depth in C.M.		Description
A	0-10	:	Pale yellow (2.5 Y 7/4) sandy loam crumb structure, slightly hard when dry, plenty root and iron concretions, gradual boundary.
B	10-25	:	Yellow (10 YR 7/6) loam, crumb structure slightly hard when dry plenty roots and iron concretions gradual boundary.

B_{1B2}	25-43	: Brownish yellow (10 YR 6/6) clayt loam, subangular blocky structure, moderately hard when dry, plenty roots and iron concretions, gradual boundary.
C	43-71	: Murrum consisting of Qyartz gravels, felspar and black iron concretions in clay matrix.

(ii) It is true that at several places, the upper layers of soil have been observed to be turncated by erosion, over which sediments from the higher adjoining region have been deposited, giving rise to different layers of soils without any genetic relation. But these sediments neither occur over the bed-rock nor are they paleoclimatic in origin. These are the most recent phenomena, erosion takes place during the period of high rainfall and run off. It truncates the upper surface of the sediments. During the decreasing stage of run - off, sediment - carrying capacity of water is reduced. Decrease in run - off favours the deposits of sediments on the turncated surface of the area. Hence, this observation also does not seem to support the concept of the 'dead erosion surface:

(iii) Morrums are present in the soil profiles. They are angular or sub-angular. They sometimes may create confusion in th recognition of the gravel beds. But they are not gravels. Had these gravel - like materials been deposited by transportation from the adjoining higher surface they would have been rounded, sub - rounded and smooth, An alternative explanation may be given that they might have travelled short distance. In any case they must have occurred on the bed - rock. But this is also not the case. We see that bed - rocks are generally covered with weathered or decomposed rocks and therefore, they cannot be explained as the product of the past climate.

(iv) It is essential to note that two types of sediments are seen in the area under study. They are alluvial sands along the lower courses of the Damodar, the Barakar, the Khudia, the Jamunia, the lijri and the Gowai, etc. and the residual soils formed from the parent rocks. These sediments are most probably the product of the most recent enviroment.

(v) The presence of mere mantle of waste and debris on the

rocks does not mean that the sculpturing of bed-rocks has totally ceased and streams of the present days flowing over superficial soil cover are spending their energy in degrading or aggrading the soil mantle. We should not forget that the decomposition of bed-rocks and formation of residual soil, is a continuous process actively observed in the Damodar catchment area and also on the Chotanagpur Planation surface.

Plate - 13 : Rocky bed of Kochi river

Numerous streams are flowing through rocky exposures. It seems that rivers have been sculpturing the bed-rock surface. The steep scarps, which are generally bare of soil mantle and streams are turbulent to eat them out. These are examples against the concept of fossil erosion surface. These are examples against conception of ceasation of erosion on bed - rocks and scarp retreat. It poses the problem "why should scarp retreat stop as long as there is steep slope and water rushes down the slopes?" (Ahmad, 1970). A beautiful example of active erosion on bed - rock may be observed at Chandro along the Konar river. The recession of slopes may also be clearly observed there.

(iv) The different planation surfaces occuring at different altitudes are the result of the successive and intermittent uplifts of the area during the Himalayan tectonic activities. These Surfaces cannot be explained as due to climatic vicissitudes in the past.

(vii) Similarly climatic parameters cannot be the guiding factors for the evolution of the scarps, waterfalls, river terraces, hot springs and tors etc. present in the area.

(viii) The faulted topography is entirely the subject matter of tectonic geomorphology and climatic or climatogenetic geomorphology cannot explain this feature at all.

(ix) No doubt, the formation of sediments, their deposition in the lowlands or at suitable places may face difficulty to be interpreted by the tectonic activities. Even then tectonic activities are there. Because the emergence and subsidence of the area due to erosion and sedimentation involve isostasy and consequently tectonic activities to maintain the balance etc.

11.8. Conclusion :

The above discussions lead us to conclude that the concept of dead erosion surface or the fossil erosion surface does not appear to be tenable in the Damodar Basin area and also in the Chotanagpur Plateau Region. Morphological features e.g. the planation surfaces, faulted trough, 'Higher rims' on scarps, in- cised meanders, river terraces, waterfalls, hot springs, tors and the magnitude of gullies find their explanation in the epeirogenetic uplifts of the area during the successive and intermittent Himalayan upheavals. These features cannot be explained by the climatic and climatogenetic Geomorphology. It seems that the supporters of climatic geomorphology underestimate the epeirovariance and petrovariance. No doubt, process has its own role to play in the evolution of landform. But no sub - aerial process can act unless tectonic activities raise the surface to be attacked by the exogenic forces. The structural features e.g. joints, faults and folds have a "widespread and significant effect on land-form evolution" (Twidale, 1971 : 2) Ahmad has remarked that "climatogenetic geomorphology had only a minor and doubtful role to play, and real work still had to be done in the field of 'tectonogenetic' geomorphology" (Ahmad, 1970 : 3). There are some substances in both these concepts. The moulding of the earth's surface is the result of the competition between exogenetic and endogenetic forces.

It means "the degrational forces acting on the earth's surface are opposed by the constructional forces" (Leopold, et al. 1964 :). These forces vary both in time and space, most probably in cyclic order. Some times endogenetic forces predominate and raise the surface of the earth above the base level. At other times the exogenetic forces predominate and gradually reduce the raised surface near to the base level. There is thus "a perpectual struggle, an uninterrupted morphological evolution, always in progress but never finished" (Tricart, 1974 : 2). Thus, morphological variation will remain as long as there is competition between the internal (tectonic) and external (climatic or environmental) forces.

12

Geomorphic Regions

The following geomorphic regions of the Damodar basin in Jharkhand (Fig. 22) have striking geomorpphological personality of their own :

1) The Inter - Plateau Ridge
2) The Khamar Pat
3) The Ranchi Planation Surface
4) The Upper Hazaribag Planation Surface
5) Giridih - Kodarma plateau
6) Bagodar Upland
7) Parasnath - Tundi Hill Area
8) The Upper Damodar Valley
9) The Lower Damodar Valley
10) The Middle Barakar Valley
11) The Lower Barakar valley
12) The Scarp Zone.

12.1. Inter Plateau Ridge :

The Inter - Plateau Ridge runs through Chandwa, Tori and Balumath with an average elevation of about 600 metres. This 'neck - like' ridge connects the Ranchi and the upper Hazaribag Planation surface. It is scarped both towards the west by the Auranga valley and towards the east by the Damodar Valley. The Auranga - Damodar divide runs through it. There are some isolated hills on it. Their elevations range from a few metres to more than 300 metres above the general level of the ridge.

It is entirely composed of the Archaean formations i.e., granite and gneiss. It forms an integral part of Ranchi and Hazaribag planation surface. It is, perhaps, a remainder of the original surface between down - faulted Auranga basin in the west and the Damodar

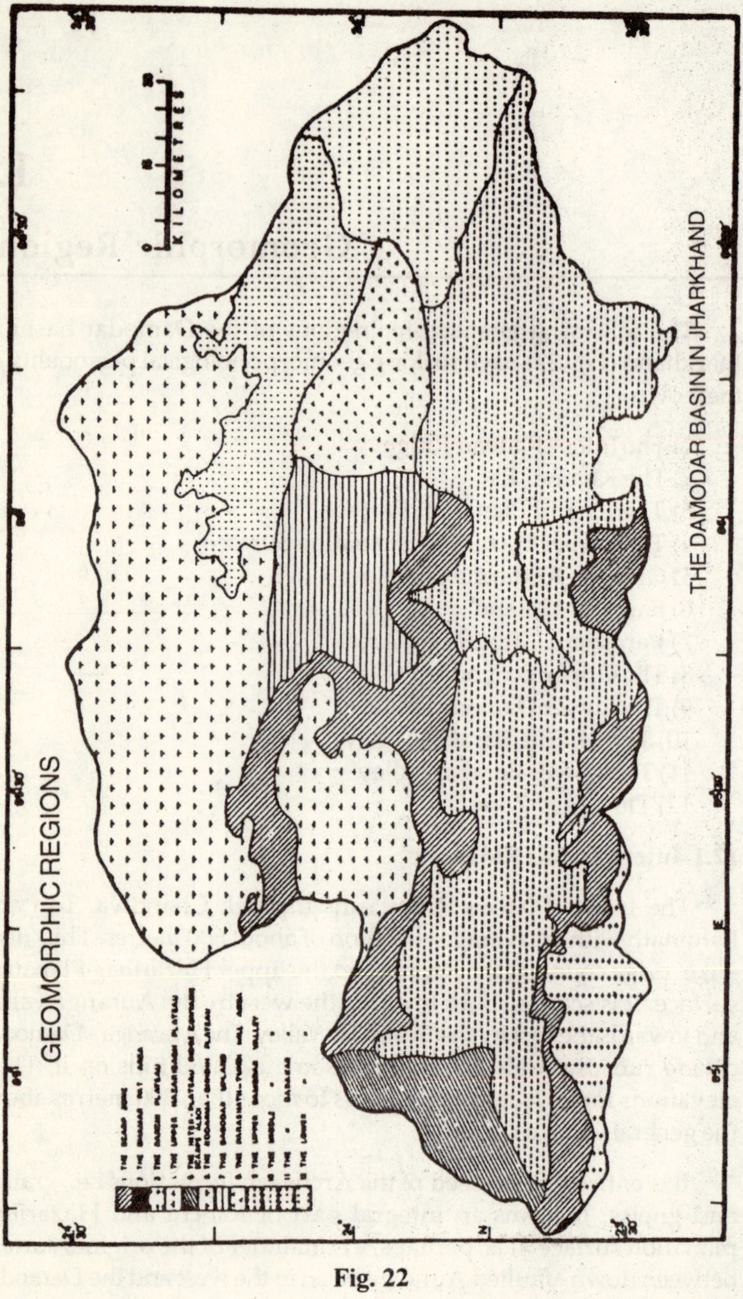
GEOMORPHIC REGIONS
THE DAMODAR BASIN IN JHARKHAND
KILOMETRES

Fig. 22

Basin in the east and, therefore, the morphological features of the ridge are in sharp contrast with those of the adjoining basins. The length of the ridge is about 40 Kms. It covers an area of 535.82 Km.2 which is 3.29 percent of the total area of the Basin.

One interesting featue is the 'Tori gap' (480 metres) between Birjangha Pahar (970 m.) and Latdag hill (1011 m.). The average gradient is about 6^0. But in the area of hills it ranges from 8^0 to 35^0 or even more.

The streams originate from both sides of the ridge. They are the tributaries of the Auranga on the west and the Damodar in the east. The scarps on either side of the ridge have been broken by the active erosion of these streams and they are receding from the east and the weat. One important point is that some streams originate from the eastern slope of Latdag hill and they turn towards west to be the tributaries of the Auranga river through Tori Gap. Here the line of divide between the Damodar and the Auranga does not run through the line of the highest elevation.

It is not easy to explain two fundamental items in the Geomorphology of the area: the inter - Plateau relatively flat surface at an elevation of about 600 metres and a number of isolated hills (locally known as tongaris) rising to 900 metres or so.

The relatively flat surface at an elevation of about 600 metres with scarped margins in the east and the west can rather be interpreted as the remnant of the peneplain. This interpretation for 600 metres surface does not appear to meet any difficulty. But occurrence on the surface, of the isolated hills as high as 900 metres or so i.e. about 300 metres above the general surface creates complications. The questions that arise (a) are these residuals, the remnant of the general peneplain which now occurs at a height of 600 metres? (b) can the work of destruction be so as to the complete wasting of relatively large portions between the residual? (c) Were the intervening rocks so weak and are the residuals so hard and resistant as compared to the general structure or lithology of the area? (d) Has so much downwasting taken place since the original peneplain took its present position? As the events belong to the past a clear answer requires much more detailed study. In the present state of knowledge, the simplest explanation that we can give is that the residual hills are relatively of hard rocks and, probably, occurred in the central 'safe' zone of divides between the adjoining streams, so that they did not undergo so much destruction as the surrounding country. The streams are still engaged in eating away side slopes and isolating the hills etc.

12.2. Pat Region :

Covering a very small area of the Basin, Khamar Pat, the highest planation surface of the Basin, is the source region of the head streams of the Damodar, the North Koel and the south koel. It makes the eastern limit of the main Pat Region of Jharkhand. The most striking featue of the region is the 'mesa - like' flattopped small plateau. The Pat has been capped with Deccan Trap which has been mostly altered to laterite. The flat surface, bounded by 1067.5 m. contours, is 1.28 Km. long and 0.32 Km. wide. The relief in this part is faint.

It has an abrupt rise from the Ranchi Planation surface (730 m.) and the Deonad Basin (530 m.). Thus, the amplitude of local relief on the northern side is greater (535 m.) than on the eastern side (336 m.). On the west a panoramic scene of dissected 'mesa' may be seen. Talsa gap (595 m.)separates the Khamar Pat - (1068 m)from Birjangha Pahar (971 m.). The average slope of the Khamar Pat area (Fig. 20) is more than 8^0. But it varies from 5^0 in the lower parts to more than 55^0 in the upper parts.

Numerous streams originate and flow down the slopes. They form radial pattern. Drainage texture is medium. The distribution and density of source heads and confluence points are medium to high. These streams are active in eating the side and pushing scarps back.

The ungraded nature of the long profiles of the streams, the narrow valleys and relatively flat interfluves clearly suggest that "the area is passing through a young stage of cycle which was started denovo after the uplift in the beginning of the Tertiary period" (Prasad, 1973 : 14). The analysis of Topographic and Hydrographic sinuosity indexes also support this conception because the percentage of Topographic Sinuosity Index is higher than the percentage of Hydrographic Sinuosity Index in this area.

Regarding the genesis of Pat Region Dunn (1942 : 12) is of the opinion that "subsequent to outpouring of Deccan trap there was prolonged erosion, and uplift to the extent of about 1000 feet (305 metres) in the western Chotanagpur". Chatterji (1946) contradicts the uplift theory of Dunn, when he proposed a 'differential erosion' theory for geomorphic interpretation of Pat Region. His observations are - either the gneissic plateau was higher in the west when it was overlain by lava flows, or difference in level, marks the extent of differential erosion, since the deposition of lava having been protected by the resistant cap, while that to the east was eroded down to the base level and subsequent rejuvenation by 610 metres would bring the lava plateaus to their present height. He further says that if we accept Dunn's uplift theory we must visualize faults along the scarps.

Ahmad (1958) raises some points against the differential erosion theory of Chatterji and supports the uplift theory of Dunn, for a geomorphic interpretation of the area. It seems reasonable to accept uplift theory for an explanation of the area. It is a peneplain uplifted during the first Himalayan upheaval.

12.3. Ranchi Planation Surface :

Ranchi Planation surface is one of the important geomorphic units of Chotanagpur Plateau. A small area of it, north of the central elevated tract (700 m.), falls under the Damodar Catchment. This elevated tract is an east - west swell by about 100 metres above the general level (610 m.) of the plateau. It acts as a line of divide between the Damodar and the Subarnrekha in the east and the Damodar and the south koel in the west. The western limit of the area is that of the foot of the Pat Region (732 m.). On the north the surface descends with steep scarps to the Damodar valley.

This plateau is "a region of flat or gently undulating country with occasional residual ridges" (Dunn, 1942 : 11). The whole surface may be divided into highlands and lowlands locally known as tanrs and dons respectively rising above which are the residual hills locally known as Tongaris. Pithouria 794 m., Chaingara 726 m., Katari 755m., Daogo 702 m., Nakta 805 m. and Harhu 687 m. are some of the notable tongaris. As a rule these tongaris west of 85^0 E. Longitudes are higher and larger than tongaris east of this line. These two levels of tongaris, probably, support the concept that the present level presents an uplifted peneplane east of the Netarhat lava region (Ahmad, 1958 : 17).

The rocks of the region are mainly granites and gneisses of Archaean era and they are exposed due to intense erosion. The relative relief (Fig. 17) is less than 50 metres in a grid of 1 mile2 (2.59 Km2). But in the area of high residual hills, it increses. Drainage texture is low to medium (Fig. 10 & 11). The density of source heads and confluence points is also low to medium . The average slope is less than 3^0 (Fig. 15), but occasionally it increases to higher value.

As noted earlier, on the north the planation surface overlooks the Damodar trough with steep slopes. "These steep slopes are almost bare rocky exposures, except at the intersection of two or more slopes along the river valleys and the scarp - foot, which offer a sojourn for the eroded material till they are removed by soil creep, land - slide, and gully erosion" (Prasad, 1965 : 47).

Now, the basic geomorphic problems are to interpret : 610 metres Ranchi Planation Surface dotted with residual hills and the steep scarp to the valley side with higher rims. The 610 metre Ranchi planation surface may be interpreted as the peneplain uplifted during the second Himalayan upheaval. That it is an uplifted surface is evident from the uniformity of the plateau surface, isolated residual hills, the high degree of truncation of structure, the rivers being indifferent to softer dharwars and harder gneisses and granites (Ahmed, 1976 : 27). There may be a number of questions as noted earlier (12.1), for geomorphic interppretation of the Tongaris. But the simplest explanation (as noted earlier) is that they are relatively of hard and resistant rocks and probably occured in the central safe zone of the divides between the adjoining streams, so that, they did not undergo so much destruction, as the surrounding country.

The genesis of scarps has been discussed earlier (Chapter : 8.1). Here, it is essential to note that the northern scarp of Ranchi Planation Surface is a fault - line scarp. the streams are active to push the scarps back, etc.

The 'Higher Rim' on the margin of the plateau is the eminence of the old mountains. This 'Rim' cannot be explained either by fault scarp or by fault - line scarp, etc.

12.4. The Upper Hazaribag Planation Surface :

The Upper Hazaribag planation surface is separated from the Ranchi planation surface by the Damodar trough. However they are connected to each other by the Inter - plateau ridge. It is similar to the Ranchi Planation Surface in morphological features except that it is more effectively dissected by the deep gullies that extend from the river banks right up to the centre of the plateau. Examples of gullies may be seen throughout the Basin.

The monotony of uplands and lowlands locally known as tarns and dons is diversified only when hills dot the landscape. Sitagarh, Bambhanbai, Silwar, Bhuswa, Banadag, Belian and kanary form a ring of seven hills round Hazaribag town. The local relief is less than 50 metres in a grid of 1 mile^2 (2.59 Km.^{2}). But in the areas of hills and deep river valley, it ranges upto 200 metre or even more.

The deep valleys of the Konar and the Bokaro have divided the plateau surface into three separate patches, arranged in an east - west elongation. It shows a panoramic view.

The upper Hazaribag Planation surface is the source region of the head streams of the Haharo, the Bokaro, the Konar, the Siwani, and the Barakar. The Mohana river and the Amanat river (outside the Damodar catchment) also originate from this plateau. The texture of drainage is coarse to medium . The density of source heads and confluence points is also low to medium.

The Upper Hazaribag plateau descends with steep scarps to the Giridih - Kodarma surface in the north and north - east, to the Bagodar upland to the east and the Damodar trough to the south. The presence of waterfalls clearly shows the break in the long profiles of the rivers.

The Upper Hazaribag plateau projects a spur towards east. A number of small hills are seen on this surface. The effects of joints in rocks on the evolution of landscape can be observed at several points. Numerous hills have been reduced to a heap of boulders, probably, due to joints in rocks.

The geomorphic problems and explanations are the same as in case of the Ranchi planation surface. But the question of the scarp which descends to 305 metres surface differs from that of the fault - line scarps on both sides of the Damodar trough. Two explanations (as noted earlier) have been advanced. The explanation, that the area was uplifted along the line where scarp is found, does not seem to be tenable as we donot visualize faulting in this section. The alternative explanation is that this scarp is 'receded scarp'. The whole of Chotanagpur Plateau was uplifted during the Himalayan Orogeny and scarp has receded inward. This view has the merit that it does not require us to prove faultings along the scarps. Secondly, the scarp is almost circular and curvilinear faults are rare anywhere in the world. One important aspect is that of mixed slopes, which are generally observed in the scarp zones and hill - sides. This can be explained due to variation in energy applied their in keeping alliance with the rejuvenation of the area.

12.5. Giridih -Kodarma Plateau :

Gridih - Kodarma Plateau extends from the Barakar - Mohana divide on the west to the kiul- Barakar and the Ajay - Barakar divides in the north - east and east. It covers an area of about 3414.41 $Km.^2$ which is 21.07 per cent of the total area of the Damodar Basin.

The western part of this region is composed of banded gneisses and granites. The area is characteristically less precipitous. The average slope is less than 3^0. The local relief is less than 50 metres. The texture of drainage is coarse. The density of source heads and confluence points is also low.

The north - western portion of this region is composed of Dharwarian rocks, mainly quatzites and schists. These rocks are frequently traversed by mica - pegmatite veins. The landscape in this section is dominated by quartzitic hills and ridges with sharp crests and almost vertical sides of bare rocks. "The schists have been eroded away and have given rise to a rolling or undulating surface" (Prasad, 1965).

A large area of this unit (from south - west to north - east) is comprised of granites and gneisses of Archaeans with intrusive character of rocks and pegmatite veins at places. The landscape is dominated by a gently undulating country with residual hills and ridge locally known as tongaris. The surface has been highly dissected by gullies.

The relative relief is less than 50 metres in a grid of 1 mile2 (92.59 $Km.^2$). But occasionally it rises upto 100 metres or even more in a grid of 1 $mile^2$ (2.59 Km^2).

The drainage texture varies from coarse to medium and density of source heads and confluence points also varies from low to medium and high .

The south - western portion of this region is drained by the upper course of the Barakar river and several tributaries join the main stream. These streams generally form dendritic pattern. They are characterized by sinuous course and V-shaped valleys with steep banks.

The basic questions are to interpret the geomorphic features e.g. the planation surfaces, the residual hills and ridges and gullies.

For an explanation of the planation surface geologists and geographers are of the opinion that it is peneplain upplifted during the last Himalayan upheaval. This is the 3rd planation surface and the above explanation for its origin does not face any difficulty the incised meander, the presence of waterfalls, potholes and distribution and development of gullies, etc. support this contention that the area is an uplifted peneplain.

Most of the residual hills in the area are of hard and resistant rocks. They are also in central ' safe ' zone of the adjoining streams. Joints in rocks have played a major role in the evolution of the hills, e.g. near Birni in Giridih district several hills have been converted into mere a heap of boulders due, probably, to joints in rocks.

Gullies, the most spectacular geomorphic features have been discussed in some detail in chapter 8 here, it is essentialto note that the magnitude of gullies can be explained, probably, due to uplift of the area, etc.

12.6. Bagodar Upland :

Bagodar Upland extends from the foot of the scarp of the Upper Hazaribagh Plateau on the west to the foot of the Parasnath hill in the east. On the north the upland is delimited by the southern boundary of the Giridih - Kodarma Plateau and on the south by the northern boundary of the Middle Damodar Valley. It covers an area of about 973.84 $Km.^2$ which is 5.99 percent of the total area of the Basin.

Most of the Bagodar Upland is composed of granites and gneisses. The gneisses are generally associated with schists. An apparent patch of gneiss may be observed in the extreme south of Bagodar anchal elongating roughly west - east and stretching up to south - westren portion of Dumari anchal and eastern part of Bishnugarh anchal. An adjacent patch of quartz - schist and quartz

- mica - schist in the north of this belt may be seen. In the middle of Bagodar anchal narrow patch of mica - schists runs from west to east.

Physiographically the area is characterized by rugged topography, long and deep river valleys, interfluvial ridges, gullies and occasionally exposed and circumdenuded hills and hillocks, etc. The presence of hills and tors has diversified the physical landscape of the area.

It is a part of the 3rd planation surface of the area. The local relief is less than 50 metres, but occasionally it rises upto 100 metres or so. The region slopes towards north, east and south - east. The average slope is less than 3^0. The texture of drainage is coarse to medium. The density of source heads and confluence points is also low to medium.

12.7. Parasnath - Tundi Hill Area :

Parasnath - Tundi Hill Area occupies 735.36 $Km.^2$ which is about 4.63 per cent of the total area of the Basin. Parasnath Hill (1366 m.), the highest point in the Basin, forms a very conspicuous landform. The topography is the result of differential erosion and "solely depends on differences of hardness of rocks. The Parasnath Hill consisting of augite enstatite granite, similar to pyroxene granites of the Nilgiri hill owes its preservation to the relative hardness of the rocks" (D.G.Manbhum, 1911 : 28). Such deep - seated instrusive bodies of hard igneous rocks like granite exposed by long continued erosion have resisted erosion with reference to the less hard gneiss of the surrounding country, and are known as residual hills.

Tundi range (742 metres) is a mere eastward extension of Parasnath with a characteristic 'Knife - edge'. Knob - like summits of many hills locally known as Munda form another type of landform. Girna Hill has a Munda, while another hill with two knoblike summits is called Do Munda (Singh, 1969 : 37).

The local relief varies from less than 100 metres to more than 700 metres in a grid of 1 $mile^2$ (2.59 $km.^2$). The highest local relief is 778 metres in a grid of 1 $mile^2$ (2.59 km^2) in the Parasnath hill area. The average slope varies from less than 4^0 in the lower parts to more than 14^0 in the upper parts. The hill-side slope is asymmetrical because the slope of the southern side of the Parasnath hill is steeper than the slope of the northern side of the hill.

Numerous streams originate from the hilly area. These streams are the agents of active erosion. The pattern of draiange is radial (Fig. 13e). Rajdaha reservoir is at the foot of the Parasnath hill.

12.8. The Upper Damodar Valley :

Covering an area of about 2074.51 $Km.^2$ which is about 12.75 % of the total area of the Basin, the Upper Damodar Valley comprises the area above 305 metres contour. The region roughly coincides with the Permo - Carboniferous trough. This part of the Damodar Valley is marked by steep and well defined scarps on the north and the south. The westren limit of the trough is that of the Inter - Plateau Ridge.

Gondwana sediments "occupy tectonic trough with faulted limits lying in a rough linear fashion and the strata dip generally towards the more faulted southern boundry "(Krishnan, 1956 : 323), sand - stones and shales with carbonaceous strata are important rocks. These rocks give rise to topographic irregularities, the sandstones forming the hills and ridges e.g. the Morcha Pahar and the Lugu Hill and shales the valley or lowlands, etc.

The Upper Damodar valley is more hilly than the surrounding gneissic planation surfaces due to differential erosion of Gondwana rocks of varying hardness. Therefore, there seems to be little differences in the terrain of the scarped slopes and the hilly areas of the valley, particularly west of 85^0 15′ E. Long. East of this line the valley is wide, and gently sloping and is sharply contrasted with the scarpland topography.

Sandstones of the Mahadeva series are hard and resistant to erosion and form bold topography e.g. Morcha Pahar (702.m) in the Karnpura basin and Lugu Pahar (977 m.) in the lower Bokaro basin. The shales and sandstones of Talchir series and Damuda series are often too soft to form prominent relief.

12.9. The Middle Damodar Valley :

The middle Damodar Valley may be described "the first step of the gradual descent from the tableland of Chotanagpur to the Delta of Bengal (D.G.1917, p.5). It extends from 305 metres contour on the west to the eastern limit of Jharkhand. The lower course of the

Damodar is in west Bengal state. This region covers an area of about 2522 Km.^{2} which is about 15.52 per cent of the total area of the Basin.

The rocks of the Middle Damodar Valley are the same as in case of the Upper Damodar valley. Sandstones and shales with good reserves of coal give rise to topographic irregularities. The igneous intrusions e.g. dykes and sills are frequently seen in the trough areas. The rolling or undulating surface is dotted here and there with isolated residual hills. The relative relief varies from less than 25 metres to more than 100 metres in a grid of 1 mile^2 (2.59 Km.^{2}). The latter figure is seen in the area of hills and ridges, locally known as Dungaris. The texture of drainage is coarse. The density of source heads and confluence points is also low. The average slope is less than 2^0.

The most striking landscape is man made hills and ridges, which is, perhaps, the universal feature in the mining areas. The heaps of boulders, deposited by the human hands may be seen in the Jharia coal fields and also in other coal fields of the Basin.

12.10. The Middle Barakar Valley :

The middle Barakar valley covers an area of 1209 Km.^{2} which is 7.44 per cent of the total area of the Basin.

It extends over the gneissic surface of the lower Hazaribag Plateau. "From the crest of the inter - fluvial ridges to the beds of the river channels the slopes are met with several breaks in the gradient, and the slopes appear like terraces which were, perhaps, produced by the phased Tertiary uplift "(Prasad, 1965 : 47). The region has been the belt of extensive gully erosion. The average slope is less than 2^0. The relative relief is less than 50 metres but it increases near hills and ridges. The texture of drainage is fine. The density of source heads and confluence points is high. The general slopes of the area is towards east.

Several isolated hills are seen to be present on projecting spurs. They owe their preservation to greater hardness of metamorphic rocks and they are in the central safe zone of divides between the adjoining streams.

The Barakar is the main stream and numerous tributaries join it from the north and the south. These streams form dendritic or tree -

like pattern. The presence of waterfall on the Usari river shows the break in the long profile of the Usari river.

12.11. The Lower Barakar Valley :

The lower Barakar valley extends over the north - eastern portion of the Basin. It covers an area of about 1194.3 $Km.^2$ which is about 7.34 per cent of the total area of the Basin. The region is composed of granites and gneisses of Archaeans and sandstones and shales with coal seams of lower Gondwana formations. In spite of the variation in geological formation the country, like the Middle Damodar Valley, is more open. The surface is dotted with residual hills. These hills owe preservation to their hard and resistant rocks.

The slope is gentle, generally less than 2^0 , but it increases near hills and ridges. The hill side slopes are steep. The relative relief is less than 50 metres but it increases near hills and ridges. The drainage density and stream frequency are low to medium. The density of source heads and confluence points are low to medium.

12.12. Scarp Zone :

Scarp zone covers an area of about 2810 $Km.^2$ or 17.28% of the total area of the Damodar Basin in Bihar. Its individuality is that it forms a continuous escarpment along the sharp edges of the planation surfaces.

The slope ranges from 5^0 to 55^0 or even more. The scarp zone assumes a moutainous complexion and precipituous look. But it has been dissected and the edges of the plateaus have been converted into a succession of sharply cut ridges and spurs. This becomes apparent when one walks in the Nalkari catchment from Pithouria to Patratu." The edges, therefore, appear as narrow fringe zone of sharply cut high hills, ridges and spurs alternating with deep valleys" (A. Prasad, 1965 : 47).

Geologically the scarp zone is composed of granites and gneisses of Archaean era. The crown of the scarp is 600 metres in the Ranchi and the Upper Hazaribag planation surfaces and 1,000 metres in the Pat region. Hills and ridges add to the general elevation of the scarps and also create complications for their geomorphic interpretations. The local relief varies from 100 metres to more than 400 metres. It is maximum in the areas of hills and ridges.

The width and the height of the scarps varies from one part to another. The northen scarp of the Ranchi planation surface is narrower and steeper than that of the scarps which bound the Hazaribag planation surface. The average width of the southern scarp is 2.5 Km. but it ranges from less than 1 Km. to 4.5 Kms. In the north the average width of the scarp is about 6 Kms. but it ranges from 2.5 Kms. to 8 Kms.

The scarps are generally continuous. But they are provided with 'gaps' or 'ghats'. These 'gaps' or 'ghats' are being used for human activities.

Numerous river valleys of varying width and depth have broken the edges of the scarps. The long profiles of several streams have been broken by the sudden break in slopes and they (streams) descend the scarp with waterfalls. Kanti waterfall on the Dhangarha river and Bokaro waterfall on the Bokaro river are some of the notable examples.

The genesis of the scarps of the Damodar Basin in Bihar has been traced in Chapter : 8.1. Here, it is essential to note that the scarps on both sides of the Damodar trough are fault - line scarps.

13

Applied Geomorphology

In the previous chapters discussions have been made at several instances to emphasize that geomorphology tells us about the shapes of hills and valleys, the degree and ferquency of slopes, the development of drainage patterns and the nature of the materials exposed in all available sections (putnam, 1962). But the role of geomorphology is not limited to this academic significance only. The truth is that geomorphology enters into our daily lives in countless ways. It has, therefore, been rightly remarked that for a better understanding of the environment of which we are a part, we must study the earth, if only to cope with, control and intelligently manage it (worcester, 1965, white, 1968. Brunsden, et al. 1978, spate, 1952, jacks and white, 1939, cooke and Doornkamp, 1974). As for example, topography obviously plays an important role in determining the most feasible high route (Thornbury, 1954). Similarly soil maps are, to a considerble degree topographic map, and the differentiation of the various members of any soil series rests fundamentally upon the different topographic conditions under which each member of the soil series develops. The problem of soil erosion is essentially a problem involving recognition and proper control of such geomorphic processes as sheetwash erosion, gullying, mass - wasting and stream erosion. Severity of erosion is not determined by the angle of slope alone, but it may be serious even on slight slopes where they are on weak materials. Application of geology to dam-site selection involves a synthesis of knowledge concerning the geomorphology, lithology and geologic structure of terrains. Many of the factors that determine the choice of a site an air-field are geomorphic or geologic in nature.

Thus, the science of geomorphology is the basis of all constructional fields right from the human settlements to terrain evaluation, military enterprise, communication and so many aspects. We noted that geomorphology has application to numerous aspects of cultural landscape. It is, of course, not manageable to include all applied aspects in the present discussion. The author has, therefore, some aspects for discussion which are noted as follows :

(1) Geomorphology and Communication

(2) Geomorphology and Hydrology

(3) Geomorphology and Settlements

(4) Geomorphology and Soil-Soil erosion and Conservation

(5) Geomorphology and Agriculture

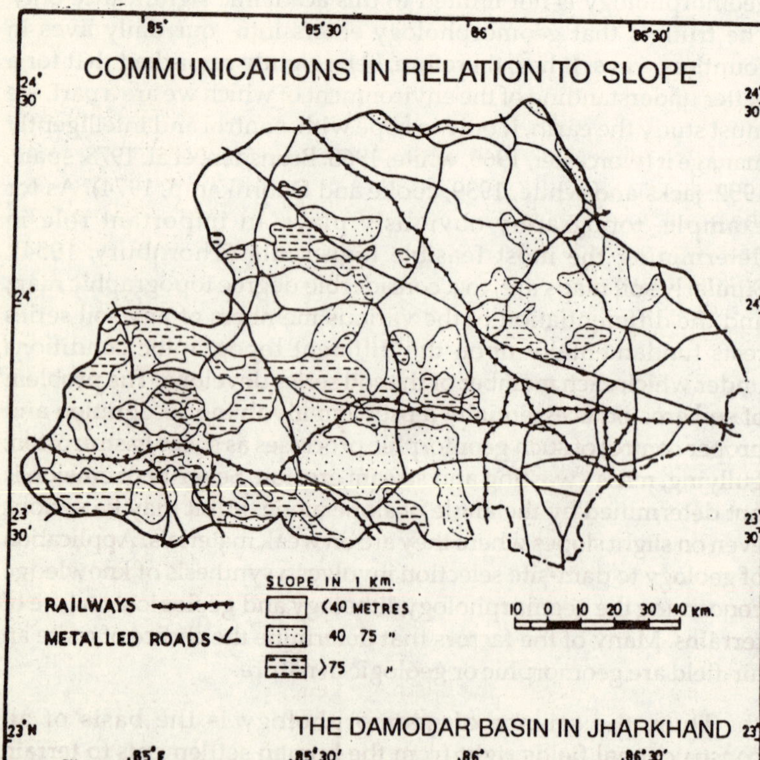

Fig. 23

13.1. Geomorphology and Communications :

Communicational lines play a major role in socio-economic development of the area. "There seem to be no other type of development (except transportation) which can affect so speedly a change in the economic and social condition" (Connor, 1965 : 21) of any particular region or country. Hence, the development of any particular area may be measured in terms of transportation as it determines the range of movement of persons and goods. "The complexities of economic interdependence manifest themselves in patterns of circulation in the economy, but circulation patterns are reflections of other things as well - types, qualities, capacities and available of transport networks, Symboisis of ubiquitous demands and specialized sources of supplies may be achieved only over the routes of available transport systems" (Berry, 1959 : 328). It is clear from the above lines : "what arteries are to a living organism, communication lines are to a region" (Pandeya, 1970 : 109). The communication system of the Damodar Basin in Jharkhand mainly include roads amd railways (Fig. 24).

The development of communication lines depends on economic, political and geomorphic factors. In the following lines, the influence of geomorphology on the pattern, development and problems of roads and railways of the Damodar Basin in Jharkhand will be anaylsed.

13.1.1. Geomorphology and Roads :

The study of geomorphological maps and road maps of the Damodar Basin in Jharkhand clearly shows the influence of geomor-phology on the pattern, development and problems of the roads. Granites, gneisses and Gondwana sandstones are found in abundance for the use of road buliding materials. Geomorphologically the Basin consists of four planation surface. They descend to their immediate lower surface with steep scarps. Accidented relief and steep scarps control the pattern and development of roads in the Basin.

The scarp zones in the Basin are generally free from roads. But they provide" geomorphic corridors" (Ahmad, 1979b : 18) e.g. 'Ghats' in the scarp zones for the development of the roads

Ranchi - Patna national highway is one of the picturesque high-

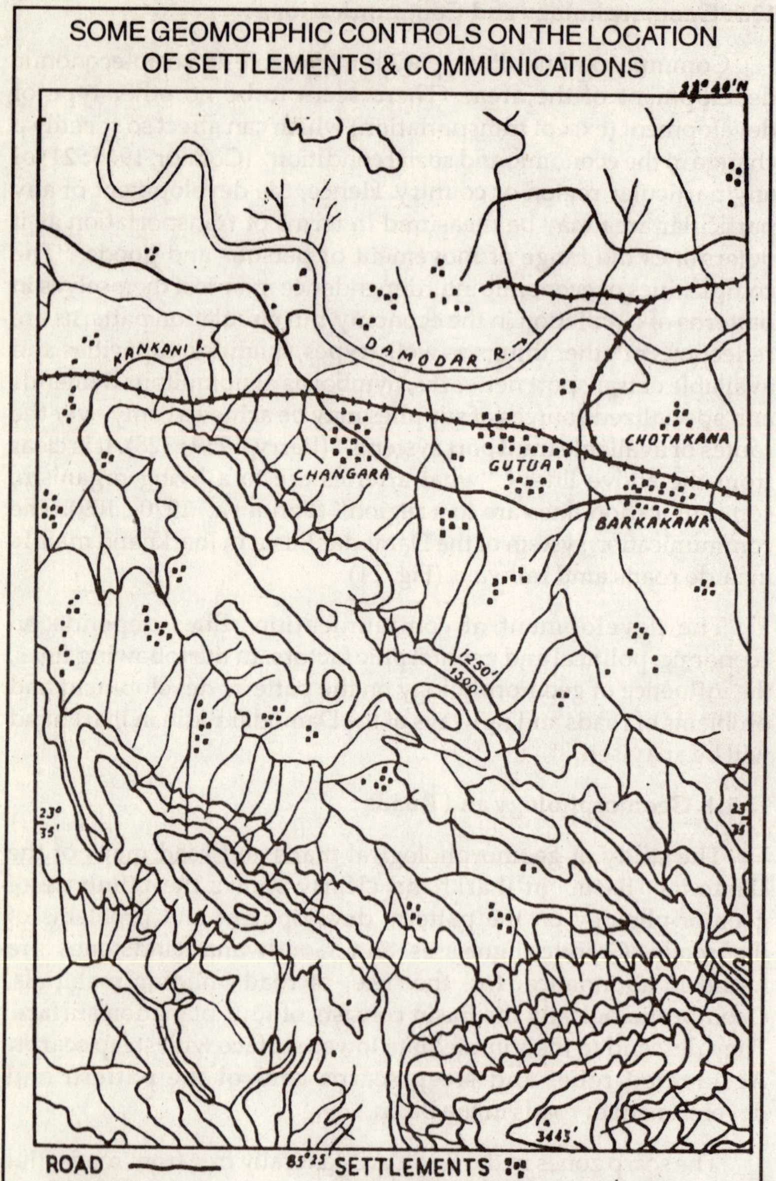
SOME GEOMORPHIC CONTROLS ON THE LOCATION
OF SETTLEMENTS & COMMUNICATIONS
23° 40'N
KANKANI R.
DAMODAR R.
CHOTAKANA
GUTUA
CHANGARA
BARKAKANA
1250'
1500'
23° 35'
23° 35'
3443'
85° 25'
ROAD
RAILWAY
RIVER
SETTLEMENTS
CONTOUR (INTERVAL 250')
1 0 1 2 3 Km.

Fig. 24

way that traverses the whole width of the Basin from Chutupalu to Kodarma. It is a narrow and winding road, through the scarp locally as 'ghat' (geomorphic corridor of Ahmad) ascending or descending the slopes of Ranchi, Hazaribag and Kodarma (Chatterji, 1969) regions. It 'meanders' to avoid hills and ridges which create a topographic hinderance for the construction of the roads. The Ranchi - Patratu road follows geomorphic corridor north of Pithouria. Similarly Dumari - Giridih and Hazaribag - Bagodar roads follow other geomorphic corridors.

Plate - 14 : National Highway through Scarp at Chutupalu

These geomorphic corridors or the passage through 'Ghats' are the critical areas of the highways. "Any vehicle failure during dark hours is frought with risks" (Ahmad, 1979b : 20). Sharp turnings cause maximum wear and tear of machinery and possibility of accidents cannot be overruled.

The Grand - Trunk Road, which acts as a spine of the urban - industrial belt, traverse the whole length (from west to east) of the Basin. It enters into the Basin at Chauparan in the upper part of the Barakar Basin. From Barkatha to Dumari it follows the divide between the Jamunia and the tributaries of the Barakar. It avoids the Parasnath - Tundi hill area in the eastern part of the Basin.

In some sections of the Basin road follows the divide line. The Ranchi - Daltonganj road over the divide between the Damodar and the south Koel from Mander to Kuru and then it follows geomorphic corridor (north of Kuru).

The road from Ramgarh to Dhanbad via Gola, Peterbar and Chas Follows the Damodar valley. The eastern and north - east -ern parts of the Basin have relatively greater lengths of roads in comparison to the western part of the Basin. The western part is more hilly than the eastern part which has greater control over the development of the roads. The road from Gobindpur to Jamshedpur via Chas provide facilities to the mining centers. The roads from Dumari to Deoghar, Giridih to Kodarma, Bagodar to Gomia and also from Bagodar to Dhanwar show the influence of the slopes on their construction.

13.1.2. Geomorpholgy and Railways :

Geomorpholgy has greater control over the pattern and development of railway than roads. It is clear from the map that roads can negotiate relatively steeper gradients. The railways are obliged to follow a certain degree of slopes. The scarp zones of the Basin are railless. The most important factor is perhaps the high gradient, sometimes more than 1 in 4, which prohibits the development of railways. The railway line from Gomoh to Daltonganj, via Gomia, Ranchi Road, Barkakana, Khelari and Tori clearly shows the influence of geomorphology on the railways. The line follows the valley flanks and it avoids the steep slopes . It is obvious from the map that the winding railway line is controlled by geomorphology (especially ruggeed topography and steep slopes) of area. The meandering pattern of west of Ranchi Road, is perhaps guided by the pettern of the Damodar river because it runs parallel to the Damodar river. It follows the foot of the southern scarp of the Basin and avoids the hills and ridge in the valley. On the west the railway line follows a geomorphic corridor i.e. the Tori gap (480 m.). This gap is flanked by Birjunga Pahar (970 m.) on the south and Latdag hill (1011 m.) on the north.

The influence of Geomorpholgy on the pattern of railway lines becomes more clear when we seen the railway running almost parallel line to the suthern scarp of the Basin. Barkakana - Muri railway line may he noted as an example of the influence of

Geomorpholgy on railway. East from Ramgarh the railway line run almost parallel to each other due to geomorpholgical control.

On the northern part of the Basin the Railway line from Chiclaki to Koderma via Parasnath, Hazaribag Road and Parsabad, and from Giridih to Maheshmunda have been constructed. The Dhanbad area has greater lenghts of railway lines than the western part of the Basin. It is also important to note that the inter - fluve (divide) between the Barakar and the Ajay is followed by a railway line.

The above discussions conclude that the railway line cannot be contructed in any direction. They are controlled mainly by the slope of the area. The higher the slope lesser development of the railways and road - ways . That is why the scarp zones of the Basin remain inaccesible to the railways. The Hazaribag town is still away from a railway due to Geomorphic reasons.

13.2. Geomorpholgy and Hydrology :

Geomorphological forms and processes have a dominant role on the hydrology of the area. It has been noted earlier that precipitation is the main source of water in the Basin. Under - ground water is limited due to geological factor. The influence of slopes on run - off is clearly observed. There is a positive relation between the velocity of water and the channel gradients. with the increase in slope of the area velocity of water increases. In the scarp zones, where the slope is relatively high, water rushes down immediately after rainfall. But change in the slope angle shows a change in the velocity of water.

Again, there is inverse relation between the degree of slope and volume of water. This is because water rushes down slope as a simple law of downflow and naturally volume of water increases with the decrease in slope. But in the Damodar Basin in Jharkhand this law is not always true. The runoff from planation surfaces accumulates to increase the volume and at the same time in the scarp zones the angle slope increases. In this case it shows positive relation.

In the Damodar Basin in Jharkhand rivers are "disconnected pools of stagnant water during the cold and hot seasons, but immediately after monsoonal showers in the hills, they come down in sudden shorlived freshets" (Ahmad, 1965 : 30). This nature of runoff is mainly due to the influence of the geomorphic from i.e. the slope.

"On the steep slopes of hills, particularly along scarps under ground water woozes out in the form of springs . Such springs are found generally at the point of contcat of two erosion surface or plane of unconformity"(Pandeya, 1970 : 13 -14) and they suggest the rejuvenation of the area.

13.2.1. Geomorpholgy and Dam - Sites :

The construction of dams and reservoirs evolves a "synthesis of knowledge concerning the geomorphology, lithology and geologic structure of terrains" (Thornbury, 1954 : 578).

A good deal about the forms and genesis of landscape of the Damodar Basin in Jharkhand has been discussed in earlier chapters. Here, it is important to note that the Damodar Basin in Jharkhand is a polycyclic region. Most of the area is composed of granites and gneisses which are stable bases for the construction of dams. Narrow outlet, watertight Basin of adequate size, and ample building materials, etc. are favourable for the construction of dams and reservoirs. The most important point to be noted is the epeirogenetic uplift of the area which facilitated the streams to be entrenched and even short embankments can empound reservoirs" Resistant residual of hard rocks are available along narrow reaches of the valley which can be joined by dams" (Ahmad, 1965 : 103).

Geomorphologically the area is, therefore, very suitable and consequently numerous dams and reservoirs have been constructed, on the main river and on its tributaries. The major dams have been located at Panchet and Tenughat on the Damodar river, the Konar on the Konar river and Tilaiay and Maithon on the Barakar river.

Apart from these major dams, several medium and minor dams have been constructed e.g. Khudia on the Khudia river, Khundoli on the Khundoli river, a tributary of the Usari, Jamunia on the Jamunia river near Bishnugarh, Patratu on the Nalkari river near Patratu etc.

Two important aspects should also be taken into consideration in the Damodar Basin in Jharkhand i.e. the siltation and faults.

The entire Damodar Basin in Jharkhand is more or less the zone of severe erosion. The deposition of silts and sands in the reservoir is reducing the lives of reservoirs. Here also the influence of geomorphic process is dominant. It creates a serious engineering problem. To ensure the long life of dams erosion should be checked or reduced immediately. No doubt, the D.V.C. is engaged in this direction but the programme should be intensified for the immediate solution of this problem. The conservation methods have been suggested elsewhere.

Another important point to be considered is that of faults. The frequency of faults can be seen on the geological maps of the Damodar trough.

The foundation of Tenughat Dam rests partly on lower Gondwana sandstones and intercalated shales and partly on crystalline metamorphics, seperated by a steeply dipping N - S fault. This fault intersects the Dam axis in a perpendicular direction on the left bank of the river and a twelve metre wide brecciated and crushed zone is associated with it. (Chowdhury, 1975 : 208). A minor fault may be observed in the south - eastern part of the Panchet Dam. Geologists are of the opinion that "the faults bounding and cutting the Gondwana rocks are old fractures which are practically dead" (Chowdhury, 1975 : 221).

In this regard we remember a very severe earthquake which occured with epicentre in Koyna nagar, on the 11th December, 1967. Geologists had declared western India as a 'stable region'. M.S. Krishnan's view that the earthquake was caused by the water of the Koyana reservior has been contradicted by E.Ahmad. He, on the basis of certain evidences, comes to the conclusion that "Koyan Nagar falls in an area which has numerous fault, some concealed below the lava, others visible within the lava region and some marginal to this tract. Therfore, the plausible conclusion about the cause of this earthquke is that it was due to the movement along some fault apparently concealed below lava" (Ahmad, 1968 :2).

Geologists are of the view that the Damodar Basin in Jharkhand is geologically stable. The area is not seismically acvtive. But we cannot ignore the fact that it is beside the seismically active belts of west bengal and North Bihar. Frequency of faults is there in the trough area. The fringe effects of earthquakes were felt in the eastern part of the Basin in 1934 during the Bihar - Nepal earth- quake. Two large dams and reservoirs, the panchet on the Damodar river and the Maithon on the Barakar river have been constructed. The eroded materials from the catchment area are being deposited. This may create isostatic imbalance in area and engineers and geologists should be watchful of it.

There may be some post - constructional problem. At Konar, a longitudinal crack was noticed along the down - stream wall, the inspection and drainage gallery of the spillway in 1963 immediately after a heavy flood was routed through the spillway and the water was stored upto the full reservior level. The crack was observed for a few years and upto 1868 and the same showed gradual widening, upto 0.50 cm. It did not show any branching. It is gathered that to seal this hair crack grounting within was done by epoxey resin (Sarma, 1975 :59). The Basin is under monsoon climate. Heavy down - pour cannot be ignored. Hence, it is essential to go in detail to know the causes of this crack and take remedial measures to avoid such incidences in future.

The above discussions conclude that the influence of geomorphology in the selection and construction of dams and resevoirs is dominant and planning should be made in consultation with, engineers, hydrologists, geologists and geomorphologists.

13.2.2. Geomorphology and Irrigation :

Geomorphological form and process have great influence on agricultural supply or irrigation. "Dissected terrain is a great Impediment in the provision of irrigation to agriculture" (Ahmad, 1979a : 5) Limited underground water and geological formations ar hinderance in digging the wells. Geomorphology of the area determines the suitability for small tanks and dams for water supply to agricultural lands in the lower parts and to the settlement in the upper parts. But during the summer seasons water table goes down and most of the tanks remain dry low land. This may be evidenced from the numerous tanks in the area constructed under minor irrigation schemes.

The control of terrains on irrigation channels is decisive. The canals are absent from the "steep scarp zones bounding the different peneplains and plateaus. In steeply sloping country or mountains, scarps or lower ridges canals are generally out of the question for a number of reasons. There is hardly such an extensive patch or level country of deep soil that needs water of the large volume involved in canals. The canals even for short lengths will have to be negotiated by numerous cuttings or tunnels and agueducts which make the canal construction prohibitive. Such terrains are also associated with the hazards of landslides or deep erosion that will seriously damage canals (Ahmad, 1971-72 : 11).

It is obvious that in the construction of canals the knowledge of geomorphological forms and processes is essential. Vaidyanathan (1971 : 319), describing the construction of the Right Bank Canal of the Tungbhadra Dam, has indicated that "with a prior knowledge of the nature of the land units perhaps the alignment of the canal might have been changed a little and thus the cost of construction considerably reduced".

Now, let us make a Geomorphological assessment of Konar irrigation canal in the area.

The Konar irrigation Project envisages the construction of a main canal and a number of distributaries to divert 49 cumes of water from Konar Reservior to irrigate 0.63 lakh hectares of cultivated and culturable land of its command area in the drought - prone districts of Hazaribag and Giridih. The main trunk canal will be open channel preceded by a tunnel to negotiate high -ground therefore.

(i) There is great geological hinderance in the construction of tunnel and canals due to granite - gneiss base of the area. It requires cutting at most places. In the construction of tunnel only harder rocks are to be faced with, because it will run underground at depth varying from 20 metres to 47 metres, while fresh and hard rocks are found at depth of only 10 metres. So, it is hard task. There is no other wayout. It was essential to negotiate high ground.

The main channels will be constructed digging in a considerable depth where mostly weathered coarse - grained rocks, sandy and red soils thereabove will have not so much obstruction in general. Therefore, the project is right in this regard.

(ii) The proposed layout of canals and distributaries takes note of the contours and slopes of the area. Generally the area slopes down towards north and east. The three branch canals, at most cases, follow spurs of main interfluvial high surface between the rivers Jamunia and Barakar in the north and Jamunia and Konar in the south. These provide high ground level for right angles of these canals, and these distributaries follow the down sloping upland (tanrs) in between the lowlands (dons) situated on either side. These sides slope down gradually on the terrains from Tanr III, II and I to Don III, II & I. This layout of the canals and distributaries will facilitate easy access to irrigation water. There is no need to lift water up, although it will have a few expectation. But after all the layout is very much according to Geomorphology.

(iii) Since the canals and distributaries run through the spurs and upland there will be no loss of good cultivated land in shape of canals. The loss of land incurred by canals and distributaries on upland will be well compensated or say over compensated by providing irrigation water to culturable waste and cultivable lands for sure, definite, regular and constant cultivation.

(iv) The vital point, that we would like to put forward and due to which we have apperhensions is the capacity of the Konar Reservoir. It is because of the marked soil erosion in the catchment of the reservoir. About 37% of the total area is under soil erosion (Table : 8 - II). Murthy and Ayengar (1956 : 19) have estimated the life of the reservoirs of the Damodar Valley Project. They have concluded that the total annual silt deposit is 331 thousand cubic metres in the Konar reservoir while it has dead storage allocation of about 72537.000 cubic metres. So, the life of the Konar reservoir is 219 years (72537000/ 331,000 = 219).

But observation reveals that the reservoir has been silting very fast and its life has beenreduced gradually. It is doubtfull if the life of reservoir will exceed to 100 or 125 years. 45 years have already been elasped and the completion of project is expected by 2000. So, we can add 5 years more. Thus, about one - half of the life of the reservoir will have elasped when diversion of water into canals will start. It is, therefore, doubful if the capacity of the reservoir will fulfill the demand of water. There is, thus, threat to irrigation potential due to the siltation in the reservoir. An apparent case has been observed at Bishnugarh. A dam of medium irrigation scheme had been constructed on the Jamunia river near Bishnugarh with a capacity of 3.91 million Kiloletres in 1954 - 55. It off - shoots two canals to irrigate 1458 hectares of land. Here is the situation that the reservoir is giving threat to the farmers year after year. It has reduced its irrigation potential due to silting in only 40 years. The farmers have been put to so much inconve-nience on accout of inadequate water supply that are led to believe that the reservoir will function some how or other for not more than 10 years to come. So, one will have to think over it while planning such a huge scheme of crores of rupees.

Concluding all these, it is suggested to arrest soil erosion in the catchment area to minimise the threat, because it is not economically feasible to remove the silt which is being deposited in the reservoir. It may be possible to reduce its quantity by soil conservation measures in the catchment area. The D.V.C. is active in this respect but the effort should be intensified.

Secondly, to meet the unfavourable consequences alternative arrangement may be thought of. Another dam and reservoir can be constructed at suitable site which linked with already constructed branch canals in furture in times of threatenings.

13.3. Geomorphology and Settlements :

The distribution, density and size of settlements are greatly influenced by geomorphology of the area. The slopes are dominant factors to influence the pattern and problems of settlements. The scarp zones of the Basin are largely negative to settlements. High gradients (25° to 45° or even more) do not offer suitable sites for settlements. Numerous streams descend the scarps. They are dotted with settlements. "At the two ends of this dotted line of settlements, i.e. on the head of the stream and at the foot of the nick - point along the scarps, comparatively large clusters are seen located. In between the foot and the crestline of the scarp a thin sprinking of settlements is found. Such settlements, generally consisting of individual huts, are located on the rounded hill tops" (Prasad, 1973 : 451).

The influence of Geomorphology on settlments is clearly seen in the upper part of the Damodar Valley. The hills and ridges e.g. Morcha Pahar (610 m.), etc. are composed of hard and resistant Gondwana sandstones. They are highly dissected by the streams. Bold topography with dissected rims and steep slopes appears to be totally devoid of settlements.

In contrast, comparatively broad, levelled and fertile area of the Damodar and its tributaries contain some excellent concentration of settlement (Fig. 24).

Gullies and ravines are problems for settlement. Villages buit along the line of divides encounter a minimum of gullying and erosion. It has been noted that in some sections of the Basin roads and railway follows the divides to avoid more costs of construction. "A number of settlements are located in the valley heads, because in the headstream region less dissection and less difficulty of terracing overbalance the disadvantage of a thinner soil cover" (Ahmad, 1976b: 235).

The influence of geomorphology is more clear in the Pat region. The Khamar Pat is a flat topped surface. It is capped by Deccan Trap and infertile lateritic soil. The run - off on the upper part is in the shape of sheet flow. Underground water is limited due to geological and topographical factors. The streams originate from the margin of the Pat and descend down the slopes. Difficult terrain, acute water problem, infertile lateritic soil and dense forest limit the number and size of settlements.

On the planation surface, the influence of geomorphology on the settlements is noteworthy. "They are situated generally on the spurs above the valley-bottom, with cultivation below in the valleys and forest and grazing grounds higher up. Another typical site is formed by ridge tops" (Ahmad, 1965 : 257).

Considering all these we conclude that on planation surfaces and other relatively level tracts "the settlements are interfluvial in location and frequently align with crestlines of divides. In the area of high hills and scarps the settlements are generally located in valleys. They are frequently seen scattered along the stream" (Prasad, 1973 : 450)

13.4. Geomorphology and Soil -Soil Erosion and Conservation :

Soil is one of the prime natural resources, uponwhich, the socio -economic development of the area is heavily based, The deterioration in the fertility of soil disturbs the economic equilibrium of the region. Thus, "to protect soil and our treasure of independence and rebuild God's greatest gift, the soil. Conservation is the new frontier" (Fink, 1952 : 39).

The soils of the Basin are residual soils being developed in situ. Deccan trap in the Khamar Pat area, sandstones and shales in the trough areas and granites and gneisses in the remaining parts of the Basin give rise to different soil types. Deccan trap gives rise to laterite or lateritic soil. Sandstones form poor sandy soil while shales yeilds more or less clayey soil. The latter is more retentive of moisture than the former. Granites and gneisses have given rise to sandy, clayey and loamy soils.

Considering the degree and character of soil formation and soil erosion we can visualize the Damodar Basin in Jharkhand as consisting of five different types of areas; the scarp zone, dissected mesa, the planation surface, the well - defined valley floor and markedly dissected surface.

In the scarp zones, the slope varies from 5° to 45° or even more. Where the slope is as steep as the latter figure, there is little question of soil formation and soil erosion. All disintergrated sediments are rapidly pulled down by gravity exposing the bare rocks. Where, however, the slope is gentler, some coarse soils may be observed. Most of the scarp zones, therefore, unless the slopes have been broken by cracks or slump, local geological structure or disposition of rocks, we can safely regard the scarp zones as hopeless zones so far soil conservation is concerned.

The Khamar Pat forms a typical geomorphic unit of its own kind. In the flatness of level it is similar to plain surface. But the region is high (1000m.) and dissected and overlooks the Damodar trough with steep scarps (gradient 1 in 4). Due to severe erosion the 'mesa' is gradually being eaten away from the sides. Soil erosion here is immaterial in the sense that it is region of infertile lateritic soil. But eroded material are being deposited in the reservoirs and hence it is a problem for engineering. It is. therefore essential to reduce the erosion.

Large area in the Damodar Basin in Jharkhand are planation surface where the average slope is less than 3°. Such areas are Ranchi and Hazaribag planation surface. Even a slope of 1° to 3° is fairly high causing soil erosion to agricultural lands. Bank erosion and gullying are conspicuous even in the dense forests, e.g. in the upper Konar and the Siwane sub - catchments, etc.

The north - eastern portion of the Basin is highly dissected surface. This is the Outer Fringe and the result of the latest phase of the rejuvenation of the area. Several geomorphological factors are combined to produce high rate of erosion. One of these factors is a formation of lateritic soil or such soil which is potentially lateritic. This is generally devoid of cementing material and subject to erosion. This belt is the zone of fine texture where drainage density and stream frequency are fairly high.

The Damodar trough has steep valley side slopes and undulating floor. Influence of geomorphology on soil formation and soil erosion is clearly seen. Soil erosion in the upper part of the valley is quite marked and higher than in Chaibasa and Ranchi planation surface, In the lower section of the valley, however, fall of gradient results in accumulation of sediments, so that in Dhanbad the soil erosion is probably the minimum in Chotanagpur.

The above discussions conclude that this nature of soil formation and soil erosion is partly due to rejuvenation of the area and partly due to different nature of rocks i.e. hard and soft rocks, etc.

There are two types of erosion : Gully Erosion and Sheet Erosion:

The extent, distribution and genesis of gully erosion have been discussed in earlier chapter. Here, it is essential to note that gullies cover about 8% of the total area of the Basin (Table : 8- II). It is eating more surface every year.

Sheet erosions may be defined as removal of uniform layer of soil or sediment from the surface of the earth. Table 8 - II reveals that 27.5% of the total area of the Basin is under sheet erosion. The Tilaiya Dam catchment has the maximum percentage (32.2%) of its total area under sheet erosion. The Khudia catchement has the minimum percentage (11.3%) of its total area under sheet erosion. Tenughat (30.7%), Konar (28.6 %), Maithon (27.9%), and Panchet (25.0%) have intermediate positions.

Two distinct processes may contribute to the removal of soil namely raindrop impact (Splash erosion) and overland flow. In overland flow the water moves as a thin film to remove the upper layer of soil or sediments.

Raindrop impact or splash erosion is "the mechanical impact of the raindrop on the weathered rock surface or soil, actually throwing up small particles in the manner of an explosive impact" (Fairbridge, 1968 : 991). The climate of the Basin is typical monsoon, with alternate dry and wet seasons. The uncertain and irregular character of monsoon sometimes brings stormy rainfall. The momentum of individual drops differs enormously and stormy rainfall is potentially more effective erosional agent than light showers. Vegetation cover reduces the effects of raindrops on soil erosion. "A complete foliage cover can entirely absorb the energy of the rain drops and prevent any dislodgement of the soil patricles : moreover, the cohesion of organically rich soils also acts as a brake on movements" (Rice, 1977 : 148). Raindrops have greater impact on the area without vegetation cover. In an experiment at Deochanda, D.V.C. (1972) has calculated the soil loss due to splash erosion (Table : 13 1).

Table : 13 -1

Monthly and Seasonal Soil Loss From the Paired Plots :

Soil Loss in lbs / acre*					
Month	Total	Loss due to surface Runoff	Loss due to splash	% of Soil loss	
	A	B	C	A	C
June	2946	658	2287	22.4	77.6
July	19652	4034	156118	22.5	79.6
August	5719	1348	4362	23.6	76.4
Sept.	5105	675	4430	13.2	86.8
Oct.	6440	691	5749	10.7	89.3
Monsoon	39862	7406	32456	18.6	

*Source: Deochanda Experiment Station : Review of Research work, Vol. 2, year 1959-69, Soil Conservation Dept. D.V.C. (Hazaribag, 1972)

It is evident from the above table that soil loss due to splash alone comes on an average to 81.4% whereas the contribution of overland flow is only 18.6%. If it is taken to be true, we can say that splash is a major problem in the Damodar Basin in Jharkhand. It is not possible to check the intensity of raindrop as it is a natural phenomenon. Therefore, the only way to conserve the soil is to protect the surface from the beating of raindrops by vegetation, grasses and other plants.

It is abvious that erosion is a problem for agriculture as it depletes in soil of its fertility. It is also a challenge for water management engineering, as it reduces the life of the reservoirs constructed for irrigation, hydel power generation and flood control, etc.

There are a number of conservation measures which have been suggested and adopted by various authorities, particularly by soil Conservations Deartment. Murthy (1970) has suggested agronomy, management of vegetation and engineering measures to check erosion. Professor Ahmad (1973) has suggested the following conservation measures:-

i) Ploughing up of incipient ravines.

ii) Straw dams.

iii) Fencing off for natural re - vegetation

iv) Breaking and levelling up of the ravines.

v) Field bunding and embodiment.

But, for satisfactory conservation schemes, it is essential to know the real and related problems of erosion, for which a thorough and up - to - date geomorphic study of the area is essential.

13.5. Geomorphology and Agriculture :

The influence of geomorphology on the pattern, development and density of agriculture is immense. In the Damodar Basin in Jharkhand different geomorphic untis e.g. the scarp zones, planation surface, accidented relief areas and valleys are characterised by different types of agriculture. "The single physical factor that broadly determines the extension as well as intensity of landuse is slope. As the slope increases, the percentage of land under plough decreases (Prasad, 1968 : 48). The percentage of land under plough in the scarp zones is insignificant due to high gradients. The slopes affect the availability of soil moisture, depth, structure and texture of soil, the amount of solar radiation, the feasibility of teracing, field pattern, possibilities of irrigation the size and distribution of holdings, continuity of culivation and the most important aspects of agriculture i.e. the crops, their nature, yields and out turn (Ahmad, 1979a : 1).

Geomorphologically the Basin consists of planaion surfaces, steep scarps and river valleyes. In the scarp zones thin layer of soil generally associated with stones and rock boulders is not suitable for agriculture. Some times crops are grown for token harvest. The remaining parts of the Basin are composed of highlands or ridges and lowlands or valleys locally known as tanrs and dons respectively. In this belt the pattern of agricultural feilds largely conforms with drainage lines.

The average slopes of the planataion surface is less than 3° It is generally 2° to 3° or even more from the bed of the channel or lowlands ('Dons') to the crest of highlands ('Tanrs'). Hence water rushes down rapidly removing the fertility of soil and soil elements. Successful cultuvation of paddy crops depends upon the flatness of the surface of land. Hence terracing "extends from the river bed to the crest upland" (Ahmad, 1979a : 3). It is important to note that on the planation surface the terraced (Ahmad, 1976a : 173) fields are relatively large but in the hilly section of the Basin the width of the terraced fields are relatively small. The pattern of terracing follows the pattern of contours in the area.

Certain crops, which require less moisture, are grown on the highlands (tanrs) and the scarp zones, without breaking the slopes of the country. maize, Gondli, Arhar and some oil seeds are grown on the slope even as high as 20° to 25° Such areas may be observed in the Nalkari Basin. Similar pattern of agriculture may be observed in the other section of the scarp zones.

It is the influence of the geomorphic forms that rice is mainly grown in the lowlands (dons) and Maize, Ragi, Pulse and oil-seed are grown on the highland area (tanrs) of the Basin.

Geomorphic forms also influence the farm irrigation which was discussed in Geomorphology and irrigation.

We have seen that the geomorphic forms e.g. the slopes, and the topography etc. control the agriculture of the area. We have also discussed the geomorphic or morphogenetic process e.g. soil erosion which are no least important in controlling the development and pattern of agriculture.

"In India more than one lac acres of good land has been going out of cultivation every year and a much larger area has been losing steadily its productive capacity due to the ravages of water and wind erosion" (Mansingh, 1959 : 57). It has been noted earlier that 8.0% (Table 8 - II) of the total area of the Basin is under severe gully erosion. Gullies are extending at the rate of 3 metres / year (Table : 8-II) A large percentage of area is under sheet erosion. It is an alarming situation and even good agriculture lands are being converted into bad lands. The areal extent of agriculture lands and agricultural productions are naturally being reduced.

Hence, not only geomorphic forms but also geomorphic processes should be taken into account for the successful agricultural planning.

Soil conservation does not mean only to protect soil from erosion but also to ensure soil fertility and soil mositure. In agricultural planning of the area therefore, geomorphic forms and processes which vitally determine soil moisture and fertility are of equally important.

Bibliography

Aghassy, J. (1970) : "Jointing, Drainage and Slopes in West African Epeirogenetic Savana Landscape", *Ann. Ass. Am. Geog;* Vol. 60, pp. 286 - 298.

Agrawal, M. (1972) : " Morphometric Analysis of some Drainage Basin of Jabalpur Plateau", *Nat. Journ. Ind;* pp. 148 - 168.

Ahmad, E. (1958) : "Geomorphic Outline of Chotanagpur" *Geog. Outlook,* Vol.2, No.3, Ranchi, pp. 16 - 22.

Ahmad, E. (1960) : "Bhusur Village : A Sample Survey of landuse Pattern and its Physical Setting in Chotanagpur", *Bombay Geographical* Magazine, pp. 45 - 52.

— (1961) : "The Rural Population of Bihar" *Geog. Rev.,* Vol. 51, Newyork, pp. 253 - 276.

— (1962): "Geomorphic Regions of Peninsular India". *Journ. of Ranchi University,* Vol.1, pp. 1 - 29.

— (1965a) : *Bihar,A Physical, Regional and Economic Geography,* Ranchi, p. 420 .

— (1965b) : "Indian Drainage", *Geog. Outlook,* vol. IV, Ranchi, pp. 9 - 21.

— (1968) : "Some Thoughts about the Origin of Koyna Nagar Earthquake", *Geog. Outlook,* Vol.5, pp. 1 - 3.

— (1969) : "Ranchi to Rajrappa : Some Geomorphic Observations", *Geog. Outlook,* Vol.6, Ranchi, pp. i -ix.

— (1970a) : "Gully Erosion in India", *21st. Inter. Geog. Cong., (1968), Selected Papers, Physical Geography, Vol.*I, *N.C.G.* Calcutta, pp. 1 - 3.

— (1970b) : *Proceedings of Symposium on Erosion Surfaces, 21st. inter. Geography Congress.* held at Ranchi on 10 - 13 Dec. 1968, pp. 3 - 13.

— (1971) : "The Ganga - A Study in River Geography", *The Geographer,* Aligarh, vol. 18, pp. 39 - 89.

— (1973) : *Soil Erosion in India,* Asia Publishing House, Bombay, p.99.

— (1976) : *Some Aspects of Indian Geography,* Central Book Depot., Allahabad, p.433

— (1978) : 'Plate Tectonics', *Lecture delivered in Bhagalpur University in Jan. 1978, under U.G.C. Visiting Professorship Programme.*

— (1979a) : "Geomorphology and Agriculture", *Five Lectures on Applied Geography,* held in Dept. of Geography, University of Rajasthan, Jaipur on 21st to 24th July, 1979, Dept. of Geog. R.U. Ranchi, pp. 1 - 9.

— (1979b(: " Geomorphology & Trade", *Five Lectures on Applied Geography,* Ibid., pp. 10 - 17.

— (1979c) : " Geomorphology and Communications", *Five Lectures on Applied Geography,* Ibid., pp. 18 - 27.

— (1979d) : " Drainage and Agriculture", *Five Lectures on Applied Geography,* Ibid., pp. 28 - 33.

& Debi, P. (1965) : "Origin of Chotanagpur Scarps", *Geog. Outlook,* Vol.4, Ranchi, pp. 81 - 85.

Ahmad, F. (1960) : " The Gondwana Basin of Sedimentation in India", *Rec. Geol. Surv. Ind.,* Vol. 86, part 3, Govt. of India, pp. 523 - 536.

(1961) : " Palaeo - Geography of the Gondwana Period in Gondwanaland", *Mem. Geol. Surv. Ind.,* Vol. 90, pp. 19 - 80.

Ahnert, F. (1970) : "Functional Relationship between Denudation, Relief and Uplift in large mid - latitude Drainage Basins", *Am. Journ. Sci.,* Vol. 268, pp. 243 - 63.

Alexander, H.S. (1962) : 'Pothole Erosion', *Journal of Geology,* Vol. 40, pp. 305 - 337.

Anantapadmanabhan,N. (1961) : "Shifting of Divides and River capture : Some examples from the Deccan", *Bombay Geog. Mag.*, Vol. VIII - IX, No. 1, pp. 105 - 111.

— (1972) : " Shifting of Divides and Stream Captures in the Western Ghats", *Symp.on Ind. Shield,* held in Mysore on 10 - 12 Dec. 1968, N.C.G. Calcutta, pp. 35 - 40.

Arogyaswami, R.N.P. (1960) : "Some Aspects of the Tectonics of the Peninsula", *Records Geol. Surv. Ind.*, Vol.89, pp. 359 - 62.

Bagchi, K. & Sengupta, A.K. (1958) : " Physiographic Regions of the Ranchi Plateau, *Bihar", Geog. Rev. Ind.,* Vol. 20, pp. 127 - 131.

Balasundram, M.S. (1976) : "General Report of the Geological Survey of India for the years, 1971 - 72", *Records Geol. Surv. India,* Govt. of India, Calcutta.

Basu, S. (1976) : " Some Fundamental Problems of Meander Formation with Special Reference to the Bhagirathi river", *Geog. Rev. of India,* Vol. 38, No. 1 (March), pp. 54 - 59.

Baulig, H. (1950a) : " William Morris Davis : Master of Methods", Ann. Ass. Am. Geogr., Vol.40, No.3, pp. 183 - 196.

(1950b) : *Essais de Geomorphologic,* publ. *de la faulte de luniversite de,* Strasbourg, Paris.

Beatty, C.F. (1958) : *Land, water and Agriculture in village communities of the Upper Damodar Valley India,* Ph.D.Thesis, University of Illinois, U.S.A. (Unpublished).

Bennet, H.H. (1955) : *Elements of soil Conservation,* Macgraw Hill Book Company, Newyork, p.993

— (1960) : " Soil Erosion in Spain", *Geog. Rev.,* Vol. 50, No.1, pp. 59 - 62.

Betal, H.R. (1970) : " Identification of Slope categories in Damodar Valley, India", *21st. Inter. Geography* Congress., 1968, Vol.1, *N.C.G.* Calcutta, pp. 4-6.

Berry, B.J.L. (1959) : " Recent Studies Concerning the Role of Transportation in Space Economy", *Ann. Ass. Am. Geography.* Vol. 49, No. 3 Sept.

Birch, P.W. (1964) : *Map and Photo Reading*, Edward Arnold (Publishers) Ltd., London, p.64

Bishop, W.W. (1958) : "Stratigraphical Geomorphology", *Essays in Geomorphology*, (Ed. by G.H. Dury), Heinemman, London, pp. 139 - 176.

Bhattacharya, A.K. (1973) : " Flood Control in the Damodar Valley", *Geog. Rev. Ind.*, vol. 35 (2), pp. 134 - 146.

Bhattacharya, D.S. & Bagchi, T.C. (1973) : *Elements of Geological Map Reading and Interpretation*, Orient Longman, New Delhi, p.78

Blackwelder, E. (1933) : " Origin of Colorado River", *Geol. Soc. Am. Bull.*, Vol.45, pp. 551 - 556.

Bose, N.K. (1940) : "Note on the Topography of Ranchi Plateau", *Cal. Geog. Rev.*, Vol.III, No.3, pp. 62 - 65.

Bose, S.C. (1945) : "Landforms, Development in the Giridih - Parasnath Region", *Cal. Geog. Rev.* Vol. VII, (Dec), pp. 106 - 111.

Bose, S.C. (1948) : *The Damodar Valley Project*, Phoenix Press Ltd. Calcutta.

— (1953) : " Natural Divisions of the Damodar Valley", *Geog. Rev. of India*, Calcutta, vol. 15, March, pp. 32 - 35.

(1972) : " Denudation Chronology of the East and West Coasts of the Deccan plateau", *Symp. on Ind. Sheild., 21st. Inter. Geog. Cong.*, held in Mysore on 10 - 12 Dec. 1968, N.C.G. Calcutta, pp. 7 - 10.

Bowman, I. (1926) : " The Analysis of Landforms : Walther Penck on the Topographic Cycles", *Geog. Rev.*, vol. 16 No.1, pp. 122 - 132.

Bretz, J.H. (1962) : " Dynamic Equilibrium and the Ozark Landforms", *Am. Journ. Sc.*, Vol. 260, pp. 427 - 438.

Brunsden, D., Doornkamp, J.C. & Jones, D.K.C. (1978) : " Applied Geomorphology : a British view ", *Geomorphology, Present Problems and Future Prospects*, Embleton, et al. (eds.) Oxford University Press, pp. 251 - 262.

Bryan, K. (1940) : " The Retreat of Slopes", *Ann. Asso. Am. Geog.* Vol. 30, pp. 254 - 268.

— (1941) : " Pre - Columbian Agriculture in the Southwest, as Conditiones byy Periods of Alluviation", *Ann. Ass. Am. Geogr.* Vol. 31, pp. 219 - 242.

Budel, J. (1959) : " Climatic Zones of the Pleistocene", *International Geol. Review,* Vol. 1(9), pp. 72 - 79.

(1968) : " Geomorphology - Principles", *The Encyclopedia of Gemorphology,* Fairbridge, (ed.), Newyork, pp. 416 - 421.

Budel, J. (1970) : " The System of Climatogenetic Geomorphology", *Proc. of Symp. on Erosion Surfaces,* held on 13 - 16 Dec. 1968, *Intr. Geog. Cong.* Ahmad and Dasgupta (eds.), *N.C.G.* Calcutta, pp. 29 - 32.

— (1973a) : " The Climatic Geomorphic System", *Climatic Geomorphology,* Derbyshire (ed.), Macmillan, London, pp. 104 - 130.

— (1973b) : " Climatogenetic Geomorphology", *Climatic Geomorphology,* Derbyshire (ed.), Macmillan, London, pp. 202 - 227.

Cain, H.R. (1976) : *Physical Geography,* Second Edition, Longman.

Calef, W. & Newcomb, R. (1953) : " An Average Slope Map, Illinois" *Ann. Ass. Am. Geogr.,* Vol.43, pp. 305 - 316.

Carlston, C.W. (1966) : " THe Effect of Climate on Drainage Density and Stream flow", *Bull. of the Inter. Assoc. of Scientific Hydrology* Vol. II, pp. 62 - 69.

Castle, F. (1966) : Five - Figure Longarithmic and other Tables, Macmillan, London,p. 64 .

Chamberlin, T.C. (1897) : The Method of Multiple Working Hypotheses", *Journ. of Geology,* Vol. 5 , pp. 837 - 848.

Chatterji, G.C. & Ghosh, P.K. (1970) : " Tectonic Framework of the Peninsular Gondwanas of India", *Records Geol. Surv. Ind.,* Vol. 98, part 2, Govt. of India, Calcutta, pp. 1 - 15.

Chatterji, S.C. (1945) : " Some Aspects of Geomorphology of the Ranchi Plateau", *Cal. Geog. Rev.,* Vol. VII, pp. 31 - 35.

Chattterji, S.C. (1946) : " Physiographic Evolution of Chotanagpur", *Cal. Geog. Rev., Calcutta,* Vol. VIII, No. 3 & 4 , pp. 39 - 42.

Chatterji, S.P. (1940) : " Gneissic Topography of Ranchi", *Cal. Geog. Rev.* Calcutta, Vol. III, pp. 45 - 48.

— (1963) : " Fifty years of Science in India', Progress of Geography, *Indian Science Congress Assoc.*, Calcutta,p. 277 .

— (1969) : *Planning Atlas of the Damodar Valley Region*, Calcutta.

Chibbar, H.L. (1953) : " The Landform, Life and Radial Drainage of Mount Parasnath, Hazaribag District, Bihar", Bull. No. 18, *Nat. Geog. Soc. od India*, Varanasi, pp. 11 - 51.

Chorley, R.J. (1957) : " Climate and Morphometry", *Journ. of Geol., Vol.* 65, pp. 638 - 68.

— (1958) : " Aspects of the Morphometry of a Polycyclic drainage Basin", *Geography. Journ.*, Vol. 124, pp. 370 - 74.

— (1962) : " Geomorphology and General Systems Theory ", *U.S. Geol. Surv. Prof. Paper*, 500.

— (1964) : " The Nodal Position and Anomalous Character of Slope Studies in Geomorphological Research", *Geog. Journ.*, Vol. 130, pp. 70 - 73.

(1965) : A Re - evaluation of the Geomorphic System of W.M. Davis", *Frontiers in Geographical Teaching*, (Methuen, London), R.J. Chorley and P. Hagget (eds.) , pp. 21 - 38.

Chorley, R.J. (1966) : " The Application of Statistical Methods to Geomorphology", *Essays in Geomorphology*, (Ed. by G.H. Dury), Heinemann, London, pp. 275 - 387.

— (1969a) : *Introduction to Fluvial Processes*, Methuen & Co. Ltd, London, p.218 .

— (1969b) : " The Drainage Basin as the Fundamental Geomorphic unit", *Intro - Flue. Proc.*, Methuen & Co. Ltd., pp. 30 - 52.

— (1969c) : *Water, Earth and Man*, Methuen & Co. London,p. 588 .

— (1978) : " Bases for theory in Geomorphology", *Geomorphology,Present problems and Future Prospects*, Embleton. et al. (eds.), Oxford University Press. pp. 1 - 13.

Beckinsale, R.P. & Dunn, A.J. (1964) : *The History of the Study of*

Landforms : Geomorphology before Davis, Vol.1, Methuen & Co. Ltd; London ,p. 678 .

— (1973) : *The History of the Study of Landforms: Life and Work of William Morris Davis,* Vol. II, Methuen & Co. Ltd., London,p. 874 . .

Chorley, R.J. & Morgan, M.A. (1962) : " Comparison of the Morphometric Features, Unaka Mountains Tennesses and N. Carolina, and Dartmoor", *Bull. Geol. Soc. Am.*, Vol. 73 pp. 17 - 34.

Chowdhury, A.K. (1975) : " Tenughat Dam Project, Bihar, " *Geol. Surv.* Ind., Miscellaneous publication No. 29, Govt. of India, Calcutta, pp. 208 - 216.

Clarke, J.I. (1958) : " Morphometry from Maps", *Essays in Geomorph.* (Ed. by G.H. Dury), Heinemann, London, pp. 235 - 274.

Cleland, H.F. (1916) : *Geology : Physical and Historical* American Book Co., Newyork,p. 655 .

Coats, D.R. (1973) : *Environmental Geomorphology,* State University of Newyork, p.262 .

Cook, R.U. & Durkamp, J.C. (1974) : *Geomorphology in Enviromental Management*, Oxford,p. 413 .

Common, R. (1958) : " Slope Failure and Morphogenetic Regions", *Essays in Geomorph.*, (Ed. by G.H. Dury), Heinemann, London, pp. 53 - 81.

Connor, A.M.O. (1965) : " New Railway Construction and the Pattern of Economic Development in East Africa", *Institute of British Geographers*, Transactions, No. 36, June.

Cotton, C.A. (1960) : *Geomorphology : An Introduction to the Study of Landforms*, Whitecombe & Tombs, Christ - Church, p. 505 .

— (1968) : " Tectonic Landscapes", *The Encyclopedia of Geomorph.*, (Ed. by R.W.Fairbridge), Reinhold Book Corporation, Newyork, pp. 1109 - 1116.

— (1969) : *Climatic Accidents in Landscape - making*. Hafner Publishing Company, Inc., Newyork, p. 354 .

— (1973) : " The Theory of Savanna Planation", *Climatic Geomor-*

phology, (Ed. by E. Derbyshire), Macmillan, London, pp. 171 - 185.

Davidson, D.A. (1976) : " Terrain Evaluation : Testing A Technique in Lower Deeside", *Scottish Geographical Magazine*, The Royal Geographical Society, Edinburg, (Sept.), pp. 108 - 119.

Davidson, R.T. (1939) : *Chambers' Twentieth Century Dictionary* (Revised and Expanded by J.L. Gaddie), W. & R. Chambers, Ltd., London, p. 1255 .

Davis, W.M. (1892) : " The Convex Profile of Bad - land Divides", *Science*, Vol.20, p. 245.

— (1899) : " The Geographical Cycle", *Geog. Journ.*, vol.14, pp. 481 - 504.

— (1904) : " Complications of the Geographical Cycle", *Reports of the VIII Inter. Geog. Congress.*, pp. 150 - 163.

— (1909) : *Geographical Essays*, Ginn & Co. Boston (Reprinted by Dover Publ. Newyork, 1954), pp. 249 - 278.

— (1913) : " Nomenclature of Surface forms on Faulted structures", *Geol. Soc. Am. Bull.*, Vol. 24, pp. 187 - 216.

— (1922) : " Peneplains and Geographical Cycles", *Bull. Geol. Soc. Am.* Vol. 23, pp. 587 - 98.

Debi, P. (1967) : *Scarps of Chotanagpur, A Study in Physiography and Human Geography*, Ranchi (Publ. in 1979).

— (1968) : " Scarps of Chotanagpur", *Geog. Outlook*, Vol. 5, Ranchi, pp. 81 - 94.

— (1970) : " The Characters of Slopes in the Scarp Zones of Chotanagpur", *Geog. Outlook*, Vol. VII, Ranchi, pp. 161 - 170.

Dickinson, G.C. (1969) : *Maps and Air Photographs*, Edward Arnold (Publishers) Ltd. London,p. 286 .

Doornkamp, J.C. & Temple, P.H. (1966) : " Surface Drainage and Tectonic instability in part of Southern Uganda", *Geog. Journ.*, *vol.* 132, part II, pp. 238 - 251.

Doornkamp, J.C. & King, C.A.M. (1961) *Numerical Analysis in Geo-*

morphology : An Introduction, Edward Arnold (Pub.) Ltd., London, 327 pp.

Dunn, J.A. (1939) : " Post Mesozoic Movements in the Northern part of the Peninsula", *Memoirs of the G.S.I.*, Vol. 73, pp. 137 - 42.

— (1942) : " The Economic Geology and Mineral Resources of Bihar Province", *Memoirs of the G.S.I.*, Vol. 78, Govt. of India, Calcutta, pp. 6 - 14.

Dury, G.H. (1951) : " Quantitative Measurement of available Relief and Texture of Dissection", *Geol. Mag.*, Vol. 88.

(1952) : " Methods of Cartographical Analysis in Geomorphological Research", *Indian Geog. Soc. Silver Jubilee Souvenir Volums*, Madras, pp. 136 - 140.

— (1960) : " Misfit Streams, Problems in Interpretation Discharge and Distribution", *Geog. Rev;* Vol.50, pp. 219 - 242.

— (1966) : *Essays in Geomorphology*, Heinemann, London.

— (1969) : " Hydraulic Geometry", *Intro. Fluv. Proc.*, R.J. Chorley (ed.), Methuean & Co. Ltd., pp. 146 - 156.

(1978) : " The Future of Geomorphology", *Geomorphology, Present Problems and Future Prospects*, Embleton, et al (eds.), Oxford University Press, pp. 263 - 274.

Dutta, G.K. (1970) : " Geomorphology : Tectniques and Applications", *Geog. Rev. of India*, Vol. 32, No.2, pp. 113 - 124.

Dutta, R.K. (1971) : " Geology of Indian Coal Measures", *Memoirs of Geol. Surv. Ind.*, Vol. 88, Govt. India.

Dutta, R.N. & Rajraman, S. (1963) : " East Bokaro Coalfield", Memoirs on Coal Exploration, Second Five Year plan, Govt. of India, Indian Bureau of Mines, Nagpur.

Dutton, C.E. (1882) : " Tertiary History of Grand Canon District", *U.S. Geol. Surv. Mem.*, Vol.2, p.76.

Embleton, C., Brunsden, D. & Jones, D.K.C. (1978) " *Geomorphology : Present problems and Future Prospects* Oxford University Press, p. 281.

Emmet, W.W. (1968) : " Gully Erosion", *The Encyclopedia of Geomorphology,* R.W.Fairbridge (ed.), Reinhold Book Corporation, Newyork, pp. 517 - 519.

Evans, p. (1971) : " The Tertiary Plateau of Bihar, " *Studies in Earth Science,* West volums, Today and Tommorrow (Pub.), New Delhi, pp. 387 - 393.

Everard, C.E. (1963) : " Contrasts in the Form and Evolution of Hill - side slopes in Central Cyprus", *Inst. Brit. Geog. Trans.,* Vol.32, pp. 31 - 47.

Fair, T.J.D. (1947) : " Slope Form and Development in the Interior of Natal", *Trans. Geol. Soc. S. Afr.,* Vol.50, pp. 105 - 119.

Fairbridge, R.W. (1968) : The Encyclopedia of Geomorphology, Reinhold Book Corporation, Newyork,p. 1295 .

Fairbridge, R.W. (1968) : " Pothole", *The Encyclopedia of Geomorphology,* Newyork, pp. 888 - 889.

Feldman, S., Harris, S.A. & Fairbridge, R.W. (1968) : "Drainage Pattern", *The Encyclopedia of Geomorphology,* Fairbridge (ed.), Newyork, pp. 285 - 291.

Fermor, L.L. (1909) : "The Mangnese - Ore Deposits of India", *Mem. Geol. Surv. Ind;* Vol. 37.

Fink, O.E. (1952) : 'Soil and Democracy' *The Land,* Vol.II, No. I.

Fox, C.S. (1930) : " The Jharia Coalfield", *Mem. Geol. Surv. of India,* Vol. 56.

— (1931) : " The Gondwana System and related Formations", *Mem. Geol. Surv. of India,* Vol. 58.

— (1934) : " The Lower Gondwana Coalfields of India", *Mem. Geol. Surv. of India,* Vol. 59.

Garnier, B.J. (1963) : *Practical Work in Geography,* Edward Arnold (Publishers) Ltd., London. p.152.

Gee, E.R. (1932) : " Geology and Coal Resources of Raniganj Coalfield", *Mem. G.S.I.,* Vol. LXI.

Gellert, J.F. (1972) : " The Climatic Morphology and Palaeoclimatic relations of the Inselberg Relief of South - west Africa." *Proc. Symp. Geomorph. Ind. shield, 21st. Inter. Geog. Cong. India (1968),* held in the dept. of Geography, Mysore University on 10 - 12 Dec., 1968, *N.C.G.*, Calcutta, pp. 51 - 54.

Ghosh, P.K. (1962) : " Recrurrence of Glaciation in Telchir Series, North Karnpura coalfield, Bihar", *Records Geol. Surv. Ind.* Vol. 87, pp. 751 - 756.

Gilbert, G.K. (1877) : "Reports on Geology of the Henry Mountains", *U.S. Geographical and Geological Survey of the Rocky Mountain Region,* Washington, D.C. Reprinted in slope Morphology, Schumm & Mosley (eds.), Dowden, Hutchinson and Ross, Inc. pp. 13 - 23.

(1880) : " Land Sculpture, Geology of the Henry Mountains" *U.S. Geographical and Geological Survey of the Rockey Mountain Region,* Reprinted in *River Morphology*, S.A. Schumm (ed.), pp. 43 - 77.

Glock, W.S. (1932) : " Available Relief as a Factor in a Profile of a landform", *Journ. of Geol.* Vol. 40, pp. 74 - 83.

Goodson, J.B. & Morris, J.A. (1962) : *A Contour Dictionary*, George G. Harrap & Company Ltd. London, Toronto, Wellington, Sydney, Fourth Edition, p. 64.

Gorrie, R.M. (1954) : *Forestry Development and Soil Conservation in the Upper Damodar Valley,* A - 15 Year Scheme, D.V.C., Calcutta, p. 62.

Gregory, K.J. & Walling, D.E. (1973) : *Drainage Basin : Form and Process,* (Arnold, London), p. 456.

Gresswell, R.K. (1972) : *Physical Geography,* Longman, p. 504.

Gupta, B.D. (1972) : " Limestone & Dolomite", *Records of Geol. Surv. Ind.,* Vol. 102, Part II Govt. of India, Calcutta.

Gupta, S.N.P. (1966) : "Geomorphology of the Upper Damodar Valley", *M.A. Dissertation (Unpublished),* Ranchi University.

——(1978 - 79) : "Gullies : A Genetic Aspect of the Drainage Basin of the Damodar", *Geog. Outlook,* Vol.14, Ranchi, pp. 77 - 86.

——(1980) : Morphological Character of the Hillside Slopes of the Damodar Basin in Bihar", *Indian Geog. Studies*, Patna.

Hack, J.T. (1960) : "Interpretation of Erosional Topography in Humid Temperate Regions", *Am. Journ. Sci.*, Vol. 258 A (Bradley Volume), pp. 80 - 97.

Harris, S.A. & Twidale, C.R. (1968) : "Geomorphic Cycles", *The Encyclopedia of Geomorphology*, R.W. Fairbridge (ed.), Reinhold Book Corporation, Newyork, pp. 237 - 240.

Heron, A.M. (1935) : "Synopsis of the Pre - Vindhyan Geology of Rajputana", *Trans. Natl. Inst. Sci. Ind.*, Vol. 1.

Holmes, A. (1975) : *"Principles of Physical Geology"*, The English Language Book Society and Nelson,p. 1288.

Holmes, C.D. (1964) : "Equilibrium in Humid - Climate Physiographic Processes," *Am. Journ. Sci.* Vol. 262, pp. 436 - 445.

Holzner, L. & Weaver, G.D. (1965) " Geographical Evolution of Climatic and Climatogenetic Geomorphology" *Ann. Assoc. Am. Geog.*, Vol. 55, pp. 592 - 602.

Horton, R.E. (1945) : "Erosional Development Streams and their Drainage Basins : Hydrophysical approach to Quantitative Morphology", *Bull. Geol. Soc. Am.*, Vol. 56, pp. 275 - 370.

Howard, A.D. (1965) : " Geomorphological System - Equilibrium and Dynamics", *Am. Journ. Sci. Vol. 263, pp. 302 - 12.*

Hubbard, G.D. (1940) : " Major Objectives of Penck and Davis in Geomorphic Studies", *Ann. Ass. Am. Geog.*, Vol.30, pp. 237 - 240.

Iyengar, S.P. (1920) : " The Acid Rocks of Mysore, A Study in Intrusion, Metamorphism and Quantitative Classification", *Bull, Mys. Geol. Dept.*, No. 9.

Jacks, G.V. & Whyte, R.O. (1939) : *The Rape of the Earth - A World Survey of Soil Erosion*, Faber, London,p. 313.

Johnson, D.W. (1933a) : " The Development of the Drainage System and the Dynamic Cycle", *Geog. Rev.*, Vol.23, No.1, pp. 114- 121.

— (1933b) : " Available Relief and Texture of Topography , A Discussion", *Journ. of Geol.*, Vol.41, pp. 293 - 305.

— (1940) : " Studies in Scientific Method: The Analytical Method of Presentation", *Journ. of Geomorphology,* Vol.3, pp. 156 - 62 and 257 - 62.

King, C.A.M. (1966) : *Techniques in Geomorphology,* Edward Arnold (Publishers) Ltd., London,p. 342.

— (1977) : " The Cyclic Principle of Physical Geography", *National Geographer,* Vol. 12, (i) Allahabad Geog. Soc., pp. 1-16.

King, L.C. (1950) : " The Study of World's Plain Lands : A new approach in Geomorphology", *Quart. Journ. Geol. Soc.,* London, pp. 101 - 131.

King L.C. (1953) : " Canons of Landscape Evolution", *Geol. Soc. Am. Bull.* Vol. 64 , pp. 721 - 752.

— (1957) : " The Uniformitarian Nature of Hill Slopes", *Transactions of the Edinburgh Geological Society,* 17, pp. 81 - 102.

— (1958) : " Correspondence on Problems of Tors", Geog. Journ., Vol.124, pp. 289 - 91.

— (1962) : *The Morphology of the Earth,* Oliver & Boyd, Edinburgh,p. 699.

Kirby, M.J. (1969) : " Erosion by Water on Hill Slopes", *Intro - Fluv. Proc.,* R.J. Chorley (Ed.), Methuen, pp. 98 - 107.

Kirk, W. (1950) : "The Damodar Valley - Valles Opima" *Geog. Rev.* Vol.40, pp. 415 - 43.

Krishanan, M.S. (1953) : "Structural and Tectonic History of India", *Mem. G.S.I.,* Vol.81, pp. 1 -109.

— (1956) : *Geology of India & Burma,* Higginbothams (Pvt.), Ltd., Madras,p. 604.

Kulkarni, D.G. & Sinha, B.N. (1969) : " A Study of Landforms in the Lower Malprabha - Krishna Doab by Polymetric Method", *The Karnatak University Journ. Science,* Vol.14, pp. 161 - 201.

Kulkarni, D.G. (1972) : "Morphological Evaluation of Landforms in Lower Malprabha - Krishna Doab", *Symp. on Ind. shield,* held in Mysore on 10 - 12 Dec. 1968, N.C.G. Calcutta, pp. 41 - 44.

Kumar, A. (1979) : *Geomorphology of Simdega and its Adjoining Area, Bihar,* Varanasi,p. 126 .

Kumar, A. & Pandey, R.N. (1975) : "Morphometric Properties of the Polygenetic Drainage Basin", *Geog. Rev. of India*, Calcutta, Vol. 3, No.4, pp. 389 - 398.

— (1980) :"Palaeoclimatic Development of Landscape in Hazaribag Plateau", *Recent Trends and Concepts in Geography,* Mandal & Sinha (eds.), Concept Publ. Co. New Delhi, pp. 251 - 363.

Lake, P. (1958) : *Physical Geography,* 4th edition, (ed. by J.A. Steers), Cambridge University Press, Cambridge,p. 483.

Langbein, W.B. & Leopold, L.B. (1966) : 'River Meanders - Theory of Minimum Variance', *U.S. Geol. Surv. Prof. paper,* 422H, p.15

Lauder, W.R. (1968) : "Stream Capture, Piracy", The Encycloped*ia of Geomorphology,* Fairbridge (ed.) Reinhold Book Corporation, Newyork, p. 1054.

Lawrence, C.R.P. (1971) : *Cartographic Methods,* Methuen, & Company p.162

Leopold, L.B. (1973) : " River Channel Change with time : An Example", *Bull. Geol. Soc. Am.,* Vol. 84, pp. 1845 - 60.

Leopold, L.B. & Miller, J.P. (1956) : "Ephemeral Streams: Hydraulic Factors and their Relation to the Drainage Net', *U.S. Geol. Surv. Profes. paper 282 - A.*

Leopold, L.B., Wolman, M.G. & Miller, J.P. (1964) : *Fluvial Processes in Geomorphology*, Eurasia Publishing House (Pvt.) Ltd. New Delhi,p. 655.

Linton, D.L. (1955) : "The Problem of Tors", *Geog. Journ;* Vol. 121, pp. 470 - 487.

Longwell, C.R. (1946) : "How old is the Colorado River", *Am. Journ. Sc.,* Vol. 244, pp. 817 - 835.

Louis, H. (1973) : " The Problem of Erosion Surfaces, Cycles of Erosion and *Climatic Geomorphology"*, Climatic Geomorphology, E. Derbyshire (ed.), Macmillan, London, pp. 153 - 170.

Luthra, J.C. (1953) : " Influence of Topography and Climate on Crop Production", *Nat. Geog. Soc. Bull.* No. 18, pp. 5 - 6.

Mackin, J.H. (1948) : " Concept of the Graded River", *Geol. Soc. Am. Bull.*, Vol. 59, pp. 463 - 572.

Martonne, E.de (1973) :"Climate : Factor of Relief", *Climatic Geography*, E. Derbyshire (ed.), Macmillan, London, pp. 61 - 75.

Mathur, L.P. & Evans, P. (1964) : "Oil in India", *XXII Intern . Geog. Cong.*, New Delhi, Esspecially pp. 9,3, and 40.

Melton, M.A. (1959) : " A Derivation of Strahler's Channel Ordering System", *Journ. Geol.*, Vol. 67, pp. 345 - 346.

Melton, M.A. (1966) : "Geomorphic Properties of Mature Dainage Systems and their Representation in an E4 Space", *Journ. of Geol.*, pp. 35 - 54.

Miller, A.A. (1935) :" The Entrenched Meanders of the Herfordshire wye", *Geog. Journ.* Vol. 85, pp. 158 - 178.

Miller, A. (1964) : *The Skin of the Earth,* Methuen and Co., London, p. 198 .

Miller, O.M. & Summersoo, C.H. (1960) : "Slope zone Maps", *Geographical Review*, Vol. 50, pp. 194 - 202.

Miller, V.C. (1953) : " A Quantitative Geomorphic Study of Drainage Basin Characterstics in the Clinch Mountain area, Virginia and Tennessee", *Project N.R. 389 - 042 Tech. Report* 3, Columbia University Dept. of Geology O. N. R. Geography Branch, Newyork.

Monkhouse, F.J. (1972) : *A Dictonary of Geography,* Edward Arnold (Publishers), Ltd. London.p. 378 .

Monkhouse, F.J. & Wilkinson, H.R. (1976) : *Maps and Diagrams,* Methuen & Co. Ltd., London, p. 432.

Monkhouse, F.J. (1977) : *Principles of Physical Geography,* Hodder and Stoughton, London,p. 570 .

Moore, W.G. (1963) : A Dictionary of Geography, Penguin Books Ltd., Middlesex., England, p. 196.

Morisawa, M.E. (1959) : "Relation of Quanitative Geomorphology to Stream flow in Representative Watersheds of the Appalachian

Plateau Province", *Project N.R. 389 - 002 Tech.* Report 20, Columbia University, Dept. of Geology, ONR, Geography branch, Newyork.

Mukhopadhyay, S.C. (1978) : "Studies on the Terrace sequence and Neotectonics of the Tista Subarnrekha Basin areas, Eastern India", *Landscapes,* Dept. of Geog. University of Burdwan, Burdwan (West Bengal), pp. 1 - 15.

— (1980) : "Some Techniques and Recent Trends in Geomorphological Studies", *Recent Trends and Concepts in Geography,* Mandal & Sinha (eds.), Concept Publishing Co., New Delhi, pp. 249 - 284.

Murthy, B.N. (1970a) : "La Sedimentation Denseless Retinues de la Damodar Valley Corporation (India)", *Commision International Des Grand Barrage* Montreal, 38, p.58. quoted by Sarkar, (1976); *Studies on the Pedogenesis of the Soils in the Upper Damodar Valley Catcment Area,* Ph.D. Thesis Unpublished, Ranchi University, Ranchi.

— (1970b) : "Effect of Soil Conservation Measures in Controlling Sediments", *Indian Journ. of Power and River Valley Development* (July), pp.

Murthy, S.R.N. (1970) : "The Breciated Quartz veins of Nurange, Hazaribag, Bihar", *Records Geol. Surv. Ind.,* Vol. 98, Part II, Govt. of India, Calcutta.

Nikiferoff, C.C. (1942) : Fundamental Formula in Soil Formation", *Am. Journ. Sci,* Vol. 240, pp. 847 - 66.

Niyogi, D. (1970) : "Morphology of the Terraces of the Subarnrekha River, India", *Selected Papers: Physical Geography,* Vol.1, Nat. Commitee for Geog. Calcutta, pp. 84 - 88.

Oldham, R.D. (1893) : "*A Manual of the Geology of India*", Govt. of India, Publication, Calcutta, Second Edition, p. 543.

Ordway, R.J. (1971) : *Earth Science,* Affiliated East - West Press Pvt. Ltd., New Delhi, p. 705.

Palmer, J. & Nelson, R.A. (1962) : "The Origin of Granite Tors, Dartmoor Devonshire", *Proc. of Yorks. Geol. Soc.,* Vol.33 pp. 315 - 340.

Pandeya, P. (1970) : *Impact of Industrialization on Urban Growth: A Case Study of Chotanagpur,* Central Book Depot, Allahabad, p. 258.

Pandey, S. (1968) : "The Evolution of Hill Slopes around Jodhpur", *The Nat. Geog. Journ. Ind.,* Vol. XIV, Part I, March, pp. 31 - 40.

Pareek, H.S. (1975) : "Geology of the area around the Panchet Dam Reservoir, District Dhanbad, Bihar, and District Purulia, West Bengal", *Records Geol. Surv. Ind.,* Vol. 106, Part 2, Govt. of India, pp. 13 - 30.

Pascoe, E.H. (1950) : *A Manual of Geology of India and Burma,* Third Edition, Vol. I, *G.S.I.* Calcutta, p. 484.

Pascoe, E.H. (1964) : *A Manual of Geology of India and Burma,* Published by *G.S.I.* Calcutta,p. 2130 .

Penck, W. (1924) : *Die Morphologirche Analyse* ein Kapitel der, Physikal, is chen, Geologic, (Engelhorns, Stuttgart), 283 pp. trans. by H. Czecj and K.C. Boswell, 1953, *Morphological Analysis of Landforms,* Mcmillan, London, p. 429 .

Pecsi, M. (1970) : "Surfaces of Planation in the Hungarian Moutains and their Bearing on Pedimentation", *Proc. of Symp. on Erosion Surfaces,* E. Ahmad and S.P. Dasgupta (eds.), pp. 33 - 39.

& Szilard, J. (1970) : "Principal Problems of Exploration and Nomenclature of Planated Surfaces", *Symposium Erosion Surfaces, 21st. Inter - Geog. Congress,* Ahmad & Das Gupta (eds.) Nat. Committee for Cong. Calcutta, pp. 17 - 28.

Pippan, T. (1968) : "Watershed", *The Encyclopedia of Geomorphology,* Fairbridge (ed.), Newyork, pp. 1222 - 1224.

Pichamuthu, C.S. (1971) : "Some Observations on the Peninsular Gneiss Complex", *Studies in Earth Science,* West volume, Today and Tomorrow, New Delhi0, pp. 82 - 97.

Pitty, A.F. (1971) : *An Introduction of the Colorado River of the West and its Tributaries,* Washington, p. 291.

Prakash, A. (1976 - 77) : "The `Pat' Region of Chotanagpur Evidence of the Origin and Evolution of the Tertiary Scarps with particular Reference to the Jashpur Pat", *Geog. Outlook,* Vol.12, Ranchi, pp. 111 - 120.

Prakash, A. (1977 - 78) : " The Laterites of Bihar and Madhya Pradesh", *Geog. Outlook,* Vol.13, Ranchi, pp. 52 - 59.

— (1979) : *Geomorphology of Lawa River Basin* Ph.D. Thesis (unpublished) p. 494 .

Prasad, A. (1965) : "The Physical Land - scape of Chotanagpur : A Study in Regional Physiography", *Geog. Outlook,* Vol.IV, Ranchi;, pp. 41 - 60.

— (1968) : "Agriculture in Chotanagpur", *Geog. Outlook,* Vol.5, Ranchi, pp. 49 - 59.

— (1973) : *Chotanagpur, Geography of Rural Settlement,* R.U. Ranchi,p. 467.

Prasad, N. (1971) : "Geomorphology in Chotanagpur", *Geography of Chotanagpur (Bihar),* Dept. of Geography, Ranchi University, pp. 1 - 9.

— (1973) : "Geomorphic Sub - Regions of the Barakar Basin", *Indian Geographical Studies Res. Bull.* No.1, Sept, Patna, pp. 50 - 57.

— (1973) : "Surajkund, the Hottest Spring of Bihar," *The Bihar Information,* Patna, March, pp. 19 - 27.

— (1974) : "The Significance of Tectonic History and Climatic changes in Geomorphic Interpretation" *Indian Geog. Studies, Res. Bull,* No.2, Patna, March, pp. 17 - 22.

— (1977) : "Stream Order Analyses of the Barakar Basin", *Indian Geog. Studies, Res. Bull.* No.9, Patna, September, pp. 70 - 73.

— (1979a) : " Hydrographic Network and Drainage Basin Analysis : A Case Study of the Barakar Basin", *Geog. Rev. of India,* Vol. 41, No. $, Calcutta, (Dec.), pp. 297 - 303.

— (1979b) : *Geomorphology of the Barakar Basin* D. Lit. Thesis, R.U. (Unpublished), p. 398.

Prasad, N.& Gupta, S.N.P. (1979) : "Stream - line Surface of Barakar Basin, India", *Indian Geog. Studies, Res. Bull.* Patna.

Prasad, N. & Sahu, S. (1980) : "Climato - genetic Geomorphology; Review and Reassessment", *Recent Trends and Concepts in Geog;* Vol. 1, Concept Publishing Company, New Delhi, pp. 385 - 405.

Prasad, N. & Singh, L.R. (1980) : "The Landscape Evolution of the Barakar Basin in the light of the Geomorphic Models", *Transactions, Inst. of Ind, Georg.*, Vol.2, No.1, (January), pp. 91 - 99.

Putnam, D.F. (1962) : "Geography is a Practical Subject", *Geography in the 20th Century*, (ed. by G. Taylor) Methuen, London.

Radhakrishna, B.P. (1964) : "Some Aspects of the Stratigraphy and Structure of the Dharwars of Mysore, South India", *Symposium on Stratigraphy, Age and Correlation of the Archaean Provinces of India, Geol. Soc. Ind. Summary of Proc. and Abstracts.*

Raisz, E. & Henery, J. (1937) : "An Average Slope Map of Southern New England", *Geog. Rev.* Vol. 27, Newyork, pp. 467 - 72.

Ram, L.N. (1968) :"Determination of Uplift Through Profile in South - East Chotanagpur", *Transactions*, Vol.5, Patna, University, pp. 117 - 126.

Roy, S. (1978) : *Studies in Geomorphology of India*, Firma KIM Private Ltd., Ganguli Street, Calcutta,p. 63.

Raza, Monis, Ahmad, A. and Mohammad, A. (1978) : *'The Valley of Kashmir: A Geographical Interpretation'* Vol. I, The Land, Vikash Publishing House, Ltd., New Delhi, p.148.

Rice, R.J. (19770 : *Fundamentals of Geomorphology*, Longman, London & Newyork,p. 387.

Roy Chowdhary, P.C. (1957) : *Bihar District Gazetteers, Hazaribag*, Secretariate Press, Bihar, Patna.

—— (1961) : *Bihar District Gazetteers, Palamau*, Secretariate Press, Patna.

— (1964) : *Bihar District Gazetteers, Palamau*, Secretariate Press, Patna.

—— (1965) : *Bihar District Gazetteers, Santal Parganas*, Secretariate Press, Patna.

Russel, I.C. (1898) : *Rivers of North America*, Newyork,p. 327.

Sarma, S.B.S. (1975) : "Damodar Valley Project, Bihar and Bengal", *G.S.I. Miscellaneous Publication*, No. 29, Govt. of India, Calcutta, pp. 50 - 61.

Satpathi, D.D.P. (1970) : "A Broad Outline of Geomorphology of Singhbhum, India", *Symposium on Erosion Surface, 21st Inter - Geog. Cong.* Ahmad & Dasgupta (eds.), pp. 68 - 70.

Satpathi, D.D.P. (72 - 73) : " Quantitative Analysis of Landforms a case Study in the Deo River Basin, Singhbhum", *Geog. Outlook*, Vol. 9, Ranchi, pp. 57 - 66.

— (1973) : *Geomorphology of Singhbhum*, D.Lit. Thesis, (Unpublished), R.U., p. 398 .

— (1975 - 76) : " Landscape Cycles of Singhbhum", *Geog. Outlook*, Vol. XI, Ranchi, pp. 27 - 35.

— (1976 - 77) : "Meander mechanism, Geometry of Meanders and Sinuosity Indexes of the major Rivers of Singhbhum, India", *Geog. Outlook*, Vol.XII, pp. 11 - 22.

Scheidegger, A.E. (1961) : "Mathematical Models of Slope Development", *Geol. Soc. Am. Bull.*, Vol. 72, pp. 37 - 49.

Schumm, S.A. (1956) : "Evolution of Drainage Systems and Slopes in Badlands at Perth Amboy, New Jersey", *Geol. Soc. Am. Bull, 67*, pp. 597 - 646.

— (1969) : "Geomorphic Implications of Climatic Changes", *Intro Fluv. Proc.*, R.J. Chorley (ed.), Methuen, London, pp. 202 - 211.

— (1972) : *River Morphology*, Dowden, Hutchinson & Ross, Inc., Stroudsburg, Pensylvania, p. 429 .

Schumm, S.A. & Mosley, M.P. (1973) : *Slope Morphology*, Dowden, Hutchinson and Ross, Inc. Stroundsburg, Pensylvania, p. 454.

Sen, P., Bhattacharya, P.B., Mukherji, R.K. & Roy, D.K. (1967) : "Evaluation of Sub - catchments in the Upper Damodar valley for undertaking Soil Conservation Measures on priority basis", *Journal of Soil and Water Conservation in India*, Vol.15, Nos. 1 & 2, (Jan - June).

Sen, S. (1966 & 67) : " River Curves and their Forms", *Indian Journal of Power and River Valley Development*, Aug. - Dec. 66 & Jan.67, Calcutta - 13, pp. 23 - 31.

Sharan, A. (1958) : "Evolution of Drainage System in the Damodar, Basin", *National Geog. Journ. Ind.*, Vol.IV, Part 3, September, pp. 131 - 136.

Sharma, S.N. (1968) : " Is the Sen Upper Sankh Line a captured Stream?". *Geog. Outlook,* Vol.5, Ranchi, pp. 42 - 48.

— (1970) : " Hanging River Basins with High Rim Walls", *Symposium On Erosion Surfaces 21st, International Geographical Congress,* held at Ranchi on 10 - 13 Dec. 1968, Ahmad & Dasgupta (eds.), *National Committee for Geography,* Calcutta, pp. 62 - 67.

— (1971) : "A case Study of Stream Piracy on the Baitarni River", *Geog. Outlook,* Vol. VII, Ranchi, pp. 59 - 70.

— (1972 -73) : " Origin of the Singhbhum Plain", *Geog. Outlook,* Vol. 9 , pp. 37 - 56.

Sharma, V.K. (1976) : " Some Hydrological Characteristics of the Damodar River", *Geog. Rev. of India,* Calcutta, Vol. 38, (Dec.), pp. 330 - 343.

Shazly, E.M.E. (1971) : " Evolution of Granitic Rocks in relation to major Tectonics", *Studies in Earth Science,* West volume, Today and Tomorrow, New Delhi, pp. 569 - 581.

Shreve, R.L. (1967) : " Infinite Topographically Random Channel Networks", *Journ. Geol.,* Vol.75, pp. 178 - 86.

Simons, D.B. (1969) : "Open Channel Flow", *Intro. Fluv. Proc.,* Chorley (ed.), Methuen & Co. Ltd., pp. 124 - 145.

Singh, G.N. (1980) : "Contributions of L.C. King in Geomorphic Thought", Recent Trends and Concepts in Geography, Mandal and Sinha (eds.), Concept Publishing Co., New Delhi, pp. 385 - 405.

Singh, O.P. (1980) : "Geomorphology of Drainage Basin in Palamau Upland", *Recent Trends and Concepts in Geography,* Mandal & Sinha (eds.), Concepts Publishing Company, New Delhi, pp. 229 - 248.

Singh, R.P. (1956) : *Geomorphological Evolution of Chotanagpur Highlands - India,* National Geographical Soc. of India, (Published in 1969), Varanasi, p. 132.

— (1957) : "Stages and Structural Growth of Chotanagpur"", *Journ. of University of Bihar,* Vol. III, No. 1, pp. 82 - 92.

— (1958) : "Structure, Drainage and Morphology of the

Chotanagpur Highlands", *Geog. Outlook,* Vol. II, No.3, Ranchi, pp. 23 - 32.

— (1970) : "Topography and Towns in the Chotanagpur Highlands", *The Magadh University Jour.* Bodh Gaya, Vol.III, No. 1, pp. 93 - 103.

Singh, R.P. & Kumar, A. (1970) : *Monograph of Bihar,* Bharati Bhawan, Patna, p.196 .

Singh, S. (1977) : "Evolution of Granito - gneissic Tors and Cut - off Mounds of Ranchi plateau *National Geographer,* Allahabad, pp. 93 - 77.

Singh, S. (1978) : " Physiographic Regions, Landforms and Erosion Surfaces of the Ranchi Plateau", *Nat. Georg.,* Vol.XIII, No. 1, pp. 43 - 65.

Singh, S. & Ghose, B. (1973) : " Inter - relationship between Quantitative Geomorphic Characteristic of the Drainage basins in Sub - humid to Humid Environment of Rajasthan", *Ann. Arid zon,* Vol.12, No. 1 & 2, pp. 82 - 99.

Singh, S., Gupta, B.D. & Kaith, D.S. (1976) : " Multiple relationship between Bifurcation Ratio and Some Morphometric Variables in Banas Catchment, Rajasthan", *Deccan Georg.,* Vol.14, No.2, pp. 151 - 156.

Singh, S. & Sharma, K.D. (1979) " Application of Aerial Photo - Interpretation in the Investigations of small Drainage Basins in the Sandstone Region of Jodhpur", *The Deccan Geog.,* Secunderabad, India, pp. 616 - 709.

Sinha, N.K.P. (1968) : *Geomorphic Evolution of Northern Rupununi Basin,* Guyana, Dept. of Geography, McGill University, Montreal, p.131.

Sinha, R.C. & Dave, V.K.S. (1963) : " Effect of Granitic Emplacement on Epidiorites of Parasnath Area, Bihar, India", *Geol. Soc. Am. Bull.,* Vol.74, No.9, (Sept.).

Small, R.J. (1972) : *The Study of Landforms,* Cambridge University Press, p. 486 .

Smeeth, W.F. (1901) : 'General Report on the work of the Department for the years 1900 and 1901', *Records Mys. Geol. Dept.,* Vol.3.

Smith, W.F. (1935) : "The Relative Relief of Ohio", *Geog. Review,* Vol.25, *The American Geog. Soc.,* Newyork, pp. 272 - 284.

Sparks, B.W. (1971) : " The Role of Rocks in Relief", *Studies in Earth Sciences,* Murthy & Rao (eds.), West volume, Today and Tomorrow, New Delhi - 5, pp. 105- 113.

— (1972) : *Geomorphology,* Longman, London,p. 371 .

Spate, O.H.K. (1952) : " Geomorphological Problems in India", *Silver Jubilee Sovenir of the Indian Geog. Soc.,* Madras, pp. 181 - 185.

— (1954) : *India and Pakistan,* Methuen, London.

Stamp, L.D. (1953) : " Land and People", *Bull. No. 18 of Nat. Geog. Soc. Ind.,* Banaras, p.5.

Starkel,L. (1965) : " Evolution of the Upper San Basin During the Neogene", *Geomorphological Problems of Carpathians,* Bratislava, pp. 127 - 170.

Stoddart, D.R. (1969a) : " Climatic Geomorphology : Review and Reassessment", *Progress in Geography,* Vol.1, Newyork.

— (1969b) : " Climatic Geomorphology", *Intro. Fluv. Proc.,* R.J. Chorley (ed.), Methuen, London, pp. 189 - 201.

Strahler, A.N. (1950) : " Equilibrium Theory of Erosional Slopes Approached by Frequency Distribution Analysis", *Amer. Journ. Sci.,* Vol. 248, Reprinted in *Slope Morphology,* Shumm & Mostey (eds.), Dowden, Hutchinson & Ross, inc. Stroudsburg, Pensylvania, pp. 161 - 201.

— (1952) : " Dynamic Basis of Geomorphology", *Bull. Geol. Soc. Am.,* Vol.63, pp. 923 - 38.

Strahler, A.N. (1964) : " Quantitative Geology of Drainage Basins and Channel Networks", *Hand Book of Applied Hydrology,* V.T. Chaw (ed.), MacGraw Hill, Newyork, Sec.4, pp. 39 - 76.

— (1968) : "Quantitative Geomorphology", *The Encyclopedia of Geomorphology,* R.W. Fairbridge (ed.), Reinhold Book Corporation, Newyork, pp. 888 - 911.

— (1971) : *Physical Geography,* Wiley Eastern (Pvt.), Ltd. New Delhi,p. 733 .

Suryanarayana, C.V. (1980) : "The Hidden Dynamism of Water", *Every man's Science*, Vol.15, No.1, Ind. Sci. Cong. Assoc., February - March, pp. 8 - 11.

Tanaka, Kitiro (1950) : "The Relief Contour Method of Representing Topography on Maps", *Geog. Rev.*, Vol.40, No.1., pp. 446 - 456.

Tanner, W.F. (1968) : " Equilibrium in Geomorphology" *The Encyclopedia of Geomorphology*, Fairbridge (ed.), Reinhold Book Corporation, Newyork, pp. 315 - 316.

Thornbury, W.D. (1954) : *Principles of Geomorphology*, John Wiley and Sons., Inc., Newyork, London, p. 618 .

Thornes, J.B. (1978) : " The Characters and Problems of Theory in Contemporary Geomorphology", *Geomorphology, Present Problems and Future Prospects*, Embleton. et al. (eds.), Oxford University Press, pp. 14 - 24.

Tricart, J. (1974) : *Structural Geomorphology*, Longman, London, p. 305 .

Tricart, J. & Cailleux, A. (1972) : *Introduction to Climatic Geomorphology* Longman, London, p. 295 .

— (1973) : "Morphogenic System and Morpho - climatic Regions" *Climatic Geomorphology*, E. Derbyshire (ed.), MacMillan, London, pp. 228 - 68.

Tuan, Y.F. (1958) : " The Misleading Antithesis of Penckian and Davisian Concepts of Slope Retreat in Waning Development", *Proc. Indiana acad. Sci.*, Vol. 67, pp. 212 - 214.

Twidale, C.R. (1968) : "Granite Landforms", *The Encyclopedia of Geomorphology*, R.W. Fairbridge (ed.), Reinhold Book Corporation, Newyork, pp. 488 - 492.

— (1971) : *Structural Landform, An Introduction to Systematic Geomorphology*, Vol.5, The MIT Press, Cambridge, Massachusetts and London, p. 287 .

— (1973) : " Some Problems of Slope Development", *Slope Morphology*, Mosley & Schumm (eds.) Dowden Hutchinson & Ross, Inc. pp. 89 - 111.

— (1976) : *Analysis of Landforms*, John Wiley & Sons Australia Pvt. Ltd.,p. 572.

— (1976a) : " Models of Landscape Evolution", *Analysis of Landforms*, John Wiley and Sons., pp. 403 - 430.

Vaidyanadhan, R. (1971) : " Studies in Applied Geomorphology"*Studies in Earth Science*, west volume,, Today & Tomorrow, New Delhi, pp. 318 - 321.

Vaidyanathan, R. & Ramana Rao, K.L.V. (1972) : " Recognition, Correlation and Age of Erosion Surface in the Iron - ore Ranges of Bihar and Orissa", in *Proc. of Symp. on Geomorphology of Indian Sheild, 21st. Inter Geog. Congress.*, held in Mysore on 10 - 12 Dec. 1968, *N.C.G.* Calcutta, pp. 33 - 34.

Verma, P. (1961) : " Diastrophic Forces and their Geomorphic Expressions in Ranchi Plateau", *The Indian Geog. Journal*, Vol. 36, No.4, Madras, (Oct. & Dec.), pp. 123 - 131.

Verma, V.K., Tandon, S.K. & Kumar, R. (1978) : *Proceedings of Symposium on Morphology and Evolution of Landforms*, Dept. of Geology, University of Delhi,p. 267.

Von Engeln, O.D. (1940) : " Symposium: Walther Penck's contribution to Geomorphology", *Ann. Ass. Am. Geog.*, Vol. 30, pp. 222 - 258.

— (1942) : *Geomorphology*, Macmillan, Newyork, p. 655 .

Wadia, D.N. (1975) : *Geology of India*, Tata McGraw - Hill Publishing Co., New Delhi, Fourth Edition, p. 508 .

Wentworth, C.K. (1930) : " Simplified Method of Determining the Average Slope of Land Surfaces", *American Journal of Science*, Series 5, Vol. 20, No. 117, pp. 184 - 194.

White, J.F. (1968) : *Study of the Earth*, Prentice Hall of India, New Delhi, p.2.

Wilson, L. (1968) : " Dynamic Geomorphology" *The Encyclopedia* of Geomorphology, Fairbridge (ed.), Newyork, pp. 297 - 301.

Wood, A. (1942) : " The Development of Hill side Slopes", *Proc. Geol. Assoc.*, Vol.53, pp. 128 - 138.

Wooldridge, S.W. (1932) : " The Cycle of Erosion and the Representation of Relief", *Scott. Geog. Mag.* Vol.48, pp. 30 - 36.

Wooldridge, S.W. & Morgan, R.S. (1959) : *An Outline of Geomorphology (The Physical Basis of Geography)*, Longman, London, p. 409.

(1959) : 'The Cycle of Erosion', *An Outline of Geomorphology*, Longman. London, pp. 159 - 172.

Worcester, P.G. (1965) : *A Text book of Geomorphology*, Affiliated East - West Press, New Delhi, p.3.

Yatsu, E. (1955) : " On the longitudinal Profile of the Graded River " Trans. Am. Geophys. Union, Vol.36 , pp. 655 - 663.

Young, A (1963) : "Some Field Observations of Slope from and Regolith and their Relation to Slope Development", *Trans inst. Brit . Geogr.*vol.32, pp. 1 - 29.

Young,R.A. & Mutchler,C.K. (1969) : " Effect of Slope Shape on Erosion and Runoff", *Trans . Amer. Soc. Agri. Eng.*, vol. 12 pp. 231 - 233, 239 (Reprinted in *slope Morphology* by Schumm & Mosley).